SpringerBriefs in Electrical and Computer Engineering

Series Editors

Woon-Seng Gan, School of Electrical and Electronic Engineering, Nanyang Technological University, Singapore, Singapore

C.-C. Jay Kuo, University of Southern California, Los Angeles, CA, USA

Thomas Fang Zheng, Research Institute of Information Technology, Tsinghua University, Beijing, China

Mauro Barni, Department of Information Engineering and Mathematics, University of Siena, Siena, Italy

SpringerBriefs present concise summaries of cutting-edge research and practical applications across a wide spectrum of fields. Featuring compact volumes of 50 to 125 pages, the series covers a range of content from professional to academic. Typical topics might include: timely report of state-of-the art analytical techniques, a bridge between new research results, as published in journal articles, and a contextual literature review, a snapshot of a hot or emerging topic, an in-depth case study or clinical example and a presentation of core concepts that students must understand in order to make independent contributions.

More information about this series at http://www.springer.com/series/10059

Dimitri Sotnik · Michael Goetz · Ivor Nissen
Editors

Cognitive Underwater Acoustic Networking Techniques

 Springer

Editors
Dimitri Sotnik
Fraunhofer Institute for Communication
Information Processing and Ergonomics
Wachtberg, Germany

Michael Goetz
Fraunhofer Institute for Communication
Information Processing and Ergonomics
Wachtberg, Germany

Ivor Nissen
Bundeswehr Technical Center for Ships
and Naval Weapons, Maritime Technology
and Research
Eckernförde, Germany

ISSN 2191-8112 ISSN 2191-8120 (electronic)
SpringerBriefs in Electrical and Computer Engineering
ISBN 978-3-662-61657-4 ISBN 978-3-662-61658-1 (eBook)
https://doi.org/10.1007/978-3-662-61658-1

This Springer imprint is published by the registered company Springer-Verlag GmbH, DE part of
Springer Nature.
The registered company address is: Heidelberger Platz 3, 14197 Berlin, Germany

Preface

This SpringerBrief is a spin-off from the EDA (European Defence Agency) research project SALSA (Smart Adaptive Long- and Short-range underwater Acoustic network), which started in November 2018. SALSA has partners from five countries, Finland, Germany, Netherlands, Norway and Sweden.

Smart underwater robots have the potential to take over lengthy and labor-intensive missions in dangerous areas. In general, the role of mobile unmanned platforms in civilian and military scenarios is becoming more important. However, crucial to their success is the seamless integration of these platforms, like Autonomous Underwater Vehicles (AUV), gliders and floats, in the wireless underwater network with surface ships, submarines, bottom/moored sensor nodes, surface gateway buoys and divers. This requires in-mission extendable and delay-tolerant underwater acoustic networks, supporting multiple frequency bands for different data rates and transmission ranges.

The SALSA project builds further on key results from the previous (EUROPA MoU) UCAC and (EDA UMS) RACUN projects to create a suitable fully-foreground (and partly-open) communication stack for mixed mobile/static networks. The aim of the project is to establish an underwater ad hoc robust acoustic network for multiple purposes with moving and stationary nodes, that can rapidly be deployed and work in littoral waters. The added value of using such systems is believed to include cost and resource effectiveness, reduced risk for personnel and ships, improved exchange of information and data, reduction of alerting times, as well as allowing forces to be self-sufficient (organic capability). This technology needs adaptive physical-layer waveforms and cognitive network strategies with underlying cooperative and non-cooperative robust processes. Today many activities require manual parameter tuning without semi-automatical processes. For the operator, the simple task of setting the source or gain levels in underwater networks is a challenge. This state-of-the-art study summarizes the current work in cognitive network-layer methods and (smart) adaptive physical-layer methods (adaptive modulation and coding, AMC) in underwater networks. The aim of this study was getting familiar with relevant research already done in this area, and finding technology gaps which are in need of a solution.

Therefore, we are glad to publish an edited version of the SALSA literature survey as this SpringerBrief. It is a sequel to the SpringerBrief Underwater Acoustic Networking Techniques from Springer 2012 [1]. Another goal is to find a common and applicable standard in the area of Underwater Internet-of-Things [ISO/IEC 30140, 30142, 30143].

Cognitive communication under water is well known by marine mammals; the evolution over millions of years has shown that this approach is successful. But the adaptive strategies are strongly intertwined with a priori and a posteriori knowledge about the environment, the scenario, the subject neighborhood and the channel conditions. The scientific question now is how to make this knowledge accessible to the machines.

The authors are affiliated with ATLAS ELEKTRONIK in Germany (Till Wiegand), with FFI in Norway (Håvard Austad, Roald Otnes, Paul van Walree), with FOI in Sweden (Arwid Komulainen), with Fraunhofer FKIE in Germany (Dimitri Sotnik, Michael Goetz) and with TNO in The Netherlands (Henry Dol, Koen Blom, Ronald in 't Velt, Ingrid Mulders) and WTD 71 in Germany (Ivor Nissen).

Chapter 1 "Overview and Definitions" was written by Dimitri Sotnik and Ivor Nissen; the chapter 2 about adaptivity at the physical layer was written by Henry Dol, Koen Blom, Paul van Walree, Roald Otnes, Håvard Austad, Till Wiegand and Dimitri Sotnik; the chapter 3 about distance estimation was written by Ivor Nissen; the chapter 4 about Delay/Disruption Tolerant Networking was written by Ronald in 't Velt, Ingrid Mulders, Arwid Komulainen and Michael Goetz; the chapter 5 about Multi Topology Routing was written by Michael Goetz and the chapter 6 "Autonomous Ad Hoc Networks" about initial contact and address assignment was written by Roald Otnes and Ivor Nissen.

The EDA SALSA project is funded by the Ministries of Defence of the five participating nations Finland, Germany, Netherlands, Norway and Sweden. This book serves the reader who is interested in this new key technology.

Particular Acknowledgments: We would like to thank the Project Arrangement Management Group of the EDA SALSA project for the permission together with Dr. Christoph Baumann, Springer Nature as Publisher, for the contracting to print this book. Susanne Netzel, WTD 71, supported it with graphics and image formatting work which is gratefully appreciated and the reviewers for the comments and remarks.

Eckernförde, Germany   Ivor Nissen

Reference

1. Otnes R, Asterjadhi A, Casari P, Goetz M, Husøy T, Nissen I, Rimstad K, Van Walree P, Zorzi M (2012) Underwater acoustic networking techniques. Springer Science & Business Media

Contents

Acronyms and Abbreviations

5G	"5th Generation", the latest generation of cellular mobile communications
ADC	Analog–Digital Converter
AMC	Adaptive Modulation and Coding
AoA	Angle-of-Arrival
APP	APPlication layer
ATLAS ELEKTRONIK	ATLAS ELEKTRONIK GmbH, Bremen, Germany
AUV	Autonomous Underwater Vehicle
BER	Bit Error Ratio
BPA	Bundle Protocol Agent
BPSK, QPSK, 8PSK, 16PSK	Phase-Shift Keying (PSK) is a digital modulation process with 2, 4, 8 or 16 symbols
CIS	Channel Impulse Response
CL	Communication Link
CRC	Cyclic Redundancy Check
CSAC	Chip-Scale Atomic Clock
CSI	Channel State Information
CTS	Clear To Send
DACAP	Distance-Aware Collision Avoidance Protocol
DFE	Decision-Feedback Equalization
DFS	Dynamic Frequency Selection
DHCP	Dynamic Host Configuration Protocol
DSSS	Direct-Sequence Spread Spectrum
DTN	Disruption Tolerant Networking
DVB	Digital Video Broadcasting
EID	Destination Endpoint Identifier
ENV	ENVironment
ESNR	Effective Signal-to-Noise Ratio
FCC	Federal Communications Commission

FEC	Forward Error Correction
FFI	Norwegian Defence Research Establishment (FFI), Kjeller, Norway
FKIE	Fraunhofer-Institut für Kommunikation, Informationsverarbeitung und Ergonomie (FKIE), Bonn, Germany
FOI	Swedish Defence Research Agency (FOI), Stockholm, Sweden
FRSS	Frequency-Repetition Spread Spectrum
FSK	Frequency Shift Keying
GPS	Global Positioning System
GSM	Global System for Mobile communications
HF	High Frequency
ISI	Inter-Symbol Interference
JANUS	JANUS is an open-source digital signaling method for underwater communications
KAM11	Kauai Acomms MURI 2011
LBL	Long BaseLine system
LF	Low Frequency
LMS	Least-Mean-Squares
MAC	Medium Access Control
MANET	Mobile Ad hoc NETwork
MCS	Modulation and Coding Scheme
MIMO	Multiple Input Multiple Output
ML	Machine Learning
MSE	Mean Square Error
MSFRSS	Multi-Stream Frequency-Repetition Spread Spectrum
MTR	Multi-Topology-Routing
MTU	Maximum Transmission Unit
NAMAC	Noise-Aware MAC protocol
NET	NETwork layer
NMEA	National Marine Electronics Association
NN	Nickname Notification
NSF	National Science Foundation
NILUS	Networked InteLligent Underwater Sensors
OFDM	Orthogonal Frequency-Division Multiplexing is a method of encoding digital data on multiple carrier frequencies
OSI	Open Systems Interconnection Model
PC	Parameter Control
PER	Packet Error Ratio
PHY	Physical layer
PI	Performance Indicator
PLL	Phase-Locked Loop

PSK	Phase-Shift Keying
QAM	Quadrature Amplitude Modulation
QEF	Quasi-Error Free
QoS	Quality-of-Service
QPSK	Quadrature Phase-Shift Keying
RACUN	Robust Acoustic Communication in Underwater Networks
RACUN-band	Frequency band 4–8 kHz
RLS	Recursive-Least-Squares
RTS	Request To Send
RX	Receive(r)
RSSI	Received Signal Strength Indicator
SALSA	Smart Adaptive Long- and Short-range underwater Acoustic network
SALSA-band	Frequency band 24–32 kHz
S-BPM	Subject-oriented Business Process Management
SBL	Short BaseLine system
SDR	Software-Defined Radio
SINR	Signal-to-Interference-plus-Noise Ratio
SNR	Signal-to-Noise Ratio
TCP	Transmission Control Protocol
TDOA	Time Difference of Arrival
TNO	Netherlands Organisation for Applied Scientific Research (TNO), The Hague, Netherlands
TOA	Time of Arrival
TPC	Transmit Power Control
TX	Transmit(ter)
UCAC	Underwater Covert Acoustic Communications
UWA	Underwater Acoustic
UWAC	Underwater Acoustic Communications
UWSN	Underwater Wireless Sensor Networks
VoIP	Voice over IP
WTD 71	Wehrtechnische Dienststelle für Schiffe und Marinewaffen, Maritime Technologie und Forschung (WTD 71), Eckernförde, Germany

Chapter 1
Overview and Definitions

Dimitri Sotnik and Ivor Nissen

Underwater networking is becoming more and more important, as it gets more interesting not only for military purposes, but for academical and industrial as well, like environment monitoring or tracking. Even the private sector has reached the underwater world with affordable underwater drones (ISO/IEC SC41 Underwater Internet-of-Things). There are a lot of different application areas, which all share the water medium. For the communication between two parties, mainly underwater acoustics is used. The propagation of the signal has similarities to wireless radio transmissions in air with multi-path effects. But the essential difference to electromagnetic waves is, that sound waves are used as a carrier with fast changing Doppler and time spread, strong absorption, and the different ambient noise situations. A sea lion and a dolphin, for example, can hear frequencies and makes sounds in different range [1]. They all have to master the challenge not to disturb each other but also to adapt the signals to the environment. While the animals already know how to manage the environmental constrains, the technical equipment still has to learn how to differentiate the signal between echoes and reflections due to multi-path propagation.

In radio communication, the progress of environmental adaptation in recent years is quite high. There are currently several studies on adaptability in the 5G network. From Big Data [2] to Machine Learning [3] as well as model-based [4] procedures are applied here. In radio links where the environmental influence is quite easy to handle, it can be estimated and even well predicted. In order to predict the influence of the environment, there are mainly a few variables to estimate the fading and the

D. Sotnik
Fraunhofer-Institut für Kommunikation, Informationsverarbeitung und Ergonomie,
Bonn, Germany

I. Nissen
Wehrtechnische Dienststelle für Schiffe und Marinewaffen,
Maritime Technologie und Forschung, Eckernförde, Germany

© The Author(s) 2020   1
D. Sotnik et al. (eds.), *Cognitive Underwater Acoustic Networking Techniques*,
SpringerBriefs in Electrical and Computer Engineering,
https://doi.org/10.1007/978-3-662-61658-1_1

path loss. For example, in the GSM band, the COST 207 committee clustered the channel impulse response into four different classes (Rural Area, Typical Urban, Bad Urban, Hilly Terrain) [5, 6].

The signal propagation underwater, however, depends on many conditions and is very difficult to estimate. It can even change during transmission, due to the impact of weather conditions, tides and of course moving communication parties. The multi-path propagation of the signal has a large effect. It is absorbed and scattered differently by the seabed, the different layers in the water and the water surface. In previous methods, the transmission was always carried out with a fixed parameter set. This had to be estimated before the tests in order to not waste energy unnecessarily and to still reach all nodes. In order to achieve optimal utilization of the crowded environment, a dynamic method, that does not waste energy must be developed for this time-variable channel. For this purpose, the SALSA project was initiated. The main objective is to enhance the RACUN network with flexibility by finding new smart processes. The resulting network should be able to extend itself autonomously with (mobile) nodes of co-operating navies, implying the need for a smart-adaptive multi-band multi-lingual delay-tolerant network. This especially concerns AUVs joining, participating in, and leaving the network, which is the central functional theme of the SALSA project. AUVs can usually communicate at high frequencies only, but may also be able to hear low frequencies. The two bands are separated in the RACUN-band (4–8 kHz used so far) and the new SALSA-band (24–32 kHz) for short/medium-range communication. Depending on various parameters, the network has to ensure an efficient and robust communication between all nodes. For this, a base for decisions is needed.

Since many theories are based on radio transmissions, this chapter will first discuss cognitive radio communication and then cover underwater aspects (similar to [6]) as basement for the adaptive waveforms at the end of the book.

1.1 Cognitive Radio in Wireless Communication

Joseph Mitola III defines cognitive motivation in [7] as follows:

> If the network wants to ask today's handsets "How many multipath components are in your location?" two problems arise. First, the network has no standard language with which to pose such a question. Second, the handset has the answer in the structure of its time-domain equalizer taps internally, but it cannot access this information. It has no computationally accessible description of its own structure. Thus, it does not "know that it knows."

Haykin described cognitive radio for wireless communication in [8]. He splits it in three main tasks: "Radio-scene analysis, which contains an estimation of inter-ference temperature [9][1] and detection of spectrum holes" [8]. The second part was

---

[1]The interference temperature has been proposed by the FCC as a metric for interference analysis. It shows the RF power available at an antenna per unit bandwidth, measured in units of Kelvin. This metric can also be converted to power flux density, makes the energy receiving in different bandwidths more comparable and is used to quantify and manage the sources of interference.

channel estimation and prediction, which contains different methods of measuring the channel condition. The third part is mainly the adaption of this configuration. The identification of spectrum holes and also the interference is in the context of underwater communication and its constantly changing conditions always just valid for this specific point in time.

In the next sections, it will be shown that these tasks can be applied to underwater communication and are also used in various studies. But there are some practical issues to discuss before.

1.2 Cognitive Radio in Underwater Communication

In difference to radio there are technological limitations, for example standard acoustic transducer cannot transmit and receive simultaneously. Some AUVs have to stop moving, if they want to receive a signal, because of propulsion noise. As the size of a transducer is proportional to wavelength, AUVs can often only use higher frequencies and stationary nodes which usually have greater transducers to transmit in lower frequencies to reach higher distances. This circumstance of different bands has so far been solved by star topologies with stationary nodes as a base station.

The article *A survey of practical issues in underwater networks* [10] highlights "a number of important practical issues that are not emphasized in the recent surveys of underwater networks, with an intended audience of researchers who are moving from radio-based terrestrial networks into underwater networks." They wrote: "Underwater networks can be characterized by their spatial coverage and by the density of nodes. These factors have significant implications for the MAC- and network-layer issues that must be addressed at design time." Thereby the authors create a taxonomy of underwater network operating regimes with the goal of providing context for the discussion later in this paper.

In the taxonomy, illustrated in Fig. 1.1, they characterize networks also on acoustical range and population. When all nodes are so close to each other that all can communicate directly, it is a single-hop network. With only a few nodes the communication can take place without much interference. However, the more nodes that are added, the more the possible throughput is limited (Dense Network). With more distributed nodes, groups emerge which are directly connected, but between the groups, often only one or a few ways exist. In this unpartitioned, multi-hop network difficulties like the hidden Terminal Problem occur and routing is needed. With a greater distance to the nodes where an acoustic connection is no longer possible and the network is only connected via mobile nodes (Ship, AUV, …) that commute between the parties, one speaks of a Delay Tolerant Network. The network must then deal with long latencies, or even with complete disruptions. When even mobile nodes could not reach the single nodes, there is no possible way to create a network.

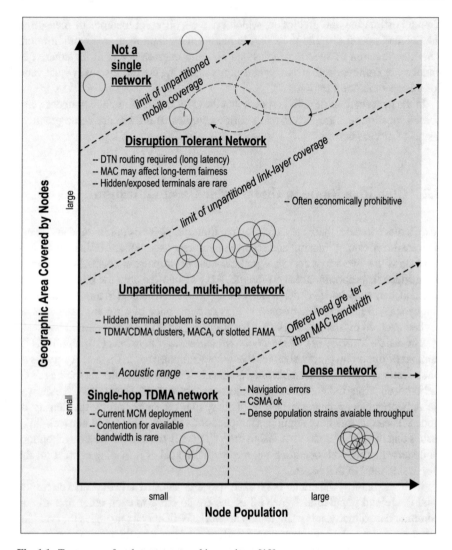

Fig. 1.1 Taxonomy of underwater networking regimes [10]

They summarized a number of practical issues differentiating underwater acoustic networks from terrestrial radio-based sensor networks. There is no single operating regime for underwater networks, instead a wide range exists. Nevertheless, the authors believed that many important underwater networks will become more mobile and more sparse than terrestrial sensor networks, with different energy and economic considerations. Underwater network protocols will have to adapt to move in sparse and dense regions, with different optimization metrics for each regime.

The Federal Communications Commission (FCC) has identified the following adaptive features, which in [6, pp. 154–156] are adapted to underwater scenarios:

- **Location Awareness** The system is able to determine the distance to the other devices operating in the same spectrum to optimize transmission parameters for increasing spectrum re-use, e.g. clock synchronization with transmitted time stamps.
- **Frequency Agility** The system is able to change its operating frequency to optimize its use in adapting to the environment.
- **Dynamic Frequency Selection (DFS)** The system senses signals from nearby transmitters/interceptors to choose an optimal operation environment. DFS lets the transmitter dynamically switch to another channel whenever a certain condition is met.
- **Adaptive Modulation** The transmission characteristics and waveforms can be reconfigured to exploit all opportunities for the usage of the spectrum. Model and/or duplex feedback knowledge is used.
- **Transmit Power Control (TPC)** The transmission power is adapted to a minimal level limit for covert communication, mammal protection, long time ocean observation, etc. TPC is a mechanism that adapts the transmission power to the communication target with the objective of having a fixed SNR level.

Since the channel introduces a long propagation delay ($\sim$6.7 s for 10 km) due to the low velocity of the acoustic carrier wave, and since there are many users using the medium, the transmission is supposed to be of very short duration. The purpose is to only occupy the channel for a short period of time to avoid the possibility of transmission collisions. The transmission therefore needs to be very efficient but also robust—this is an optimization task with many parameters.

In chapter "Adaptivity at the Physical Layer" the physical waveforms have to support different parameter sets (profiles), including parameters from:

Channel Coding: The data source as concatenation of

(a) user bits (application layer),
(b) network bits (network layer) and
(c) a small amount of user-configurable, physical parameters like the total number of transferred bits

need to be channel-coded with the goal to a desired robustness level.
Common codes in underwater acoustic communications (UWAC) are convolutional codes [11, 12], BCH codes [13], turbo codes [14–16], LDPC codes [17, 18] or polar codes [19]. Small amounts of info bits are not optimal for achieving high coding gains with powerful codes like LDPC or turbo codes, which usually require a longer data stream as input for good performance. This results in even longer output vectors, which require long interleavers. The channel decoding is then usually done in an iterative decoder where soft information is exchanged between the constituent decoders or the decoder and the equalizer (in case of

turbo equalization) [11]. The code is one of the things that can be adjusted in the parameter set.

Mapping: For high spectral efficiency, a higher-order modulation scheme is desirable, like phase shift keying variants (BPSK up to 8PSK) or, with restrictions, quadrature amplitude modulation techniques (16QAM). Clearly, the larger the constellation size, the more the Bit-Error-Ratio (BER) will rise since the constellation points get closer together and noise will cause more erroneous decisions. Most common is using a QPSK mapping for underwater communications, e.g. in [13, 20]. But also results for higher-order mappings were published, like 16QAM in [18] or up to 64QAM in [17].

Modulation: For best usage of the available bandwidth with high spectral containment, multicarrier systems, especially in the form of orthogonal frequency-division multiplexing (OFDM), are widely used. These were already presented in UWAC, e.g. early described for long-range in [12] and in [13, 17, 20, 21]. An important issue with OFDM is to maintain the orthogonality of the subcarriers which will be corrupted by Doppler effects through the channel. Therefore, methods for Doppler estimation and compensation as in [22] or [23] have to be employed. The single-carrier multi-band modulation FRSS is a good candidate for adaptive modulation, here the number of frequency (sub)bands and the bandwidth are important parameters. The general modulation equation is given in (1.1), multi-codes, multi-carrier and multi-subsymbols together with the shaping forms the time signal of one transmit symbol [24].

$$s_\nu(t) = \mathrm{Re}\left(\sum_{\kappa=0}^{N_{\text{Codes}}-1} \sum_{\mu=0}^{N_{\text{Carriers}}-1} \sum_{\eta=0}^{N_{\text{Subsymbols}}-1} c_{\nu,\mu,\eta,\kappa}\, g(t - (\nu N_{\text{Subsymbols}} + \eta)T_S)e^{2\pi l f_{\mu,\nu} t} \right) \quad (1.1)$$
$$\nu = 0, 1, 2 \ldots$$

Channel Estimation and Equalization: The main purpose of a preamble is synchronization and indicating the start of the data block. Another purpose of the detection preamble could be to allow initial channel estimation and Doppler compensation. Channel equalization is most often done by employing decision-feedback equalization (DFE) which was first reported for coherent underwater communications in [25]. Other authors presented underwater modem designs with iterative equalization [14, 16, 26]. Since the channel is time-variant, the Channel State Information (CSI) and the equalizer taps need to be updated regularly. Hence, the channel usually needs to be tracked. Besides the decision-directed adaptation, regular updates through the continuous evaluation of training symbols is possible. This, in turn, reduces the effective data rate through additional redundancy. One important parameter is the structure of the equalizer taps internally, as Joseph Mitola III explains in the beginning of this chapter.

At the end, we can split it into:

- **TX parameters**: Source level (transmission power) and the profile (no. of redundant bands, Mapping (constellation size), Code rate, Training lengths for the pilots, PHY message length (or number of blocks of e.g. 128 Bits)
- **RX parameters**: Hydrophone gain level, Equalizer parameters (tap-update method, filter lengths, PLL settings, signal-to-noise-ratio, signal-to-reverbration-ratio).
Other parameters are discussed in Chapter 2.

Nodes in underwater networks, i.e. an AUV that drives towards or away, have to configure their modem automatically in order to transmit the message in the best possible way. For this a basis is needed which makes it possible to make a correct decision. The best place to make decisions is in the Network Layer, as it has the knowledge about the nodes, the links and its capabilities. At the moment, the technical status is poor. Many researchers are working on methods to achieve adequate adaptability. They either adjust the signal directly at reception to be robustly, or make an adjustment based on previous messages to get the right configuration before the next transmission begins. Examples are:

Chapter 3: It is not only necessary to adjust parameters for an incoming signal, it is also important to adjust the parameters for sending a signal. With a reasonable distance estimation, the AUV could adjust, for example, the source level to reach only the preferred communication partner. Further the decision between lower and higher frequency bands is as well dependent on the distance.

Chapter 4: But the AUV could also have participated in the network before, and might have to adjust its configuration again.

Chapter 6: On a "higher level" of adaptivity, for example, a base communication technique should be arranged on a first occurrence of a partner. Both have to ensure that the message is understood by the peer. Also, a new member in the network has to adapt to the currently used capabilities.

All these cases have to be specified into a uniform set of rules, with small processes with effectiveness as the highest goal. This code of conduct [27] has the advantage of knowing not only information at the own node, it is also possible to derive it from the possible reactions, as they have the same rules to be followed [28]. The measured values and information required for this can be derived from two different sources. A distinction is made between Inter-Knowledge and Intra-Knowledge.

1.3 Sonar Equation

In a simple underwater communication setup, the system needs various types of information prior to communicating. But only a few of them are available, or have to be modeled before use. The sonar equation (1.2) represents the complete basis of this link. It defines the relation between the source level used and the minimum detection threshold needed, considering all the loss factors. SE is the signal excess, SL is the source transmit level [dB re μPa@m], TL is the transmission/propagation loss [dB@m], NL is the noise level [dB re μPa], DI is the directivity index [dB] and DT is the detection/decode threshold [dB] [29, 30]:

$$SE = SL - TL - NL + DI \geq DT \qquad (1.2)$$

The information for the sonar equation is either gained as input or used as output. System depth and location, for example, are input values for the transmission range to derive output from it. This equation is the incoherent model for the estimation of SL, gain, NL (ambient and self noise) and SNR. Additional parameters are the time and Doppler spread. The a priori information can be separated into four categories based on their nature of availability to a cognitive system. The values are determined by the waveform receiver, and controlled by the network layer. Based on [30] we have:

Information from a environmental database
  E.g., temperature and/or salinity over pressure, bathymetry …
Information from a scenario
  E.g., Number of Sources (Tx/Rx), Source Depth,Time period …
Information from environmental measurement
  E.g., Noise Level, SNR derived from SE, Speed of Sound …
Information from a system model
  E.g., delay time, Source level, Pilot pattern …

The main challenge is the gap in the knowledge base. It is a big question, could a concept like [30] with in-situ raytracer model and environment database work with trustworthiness?

1.4 Intra-/Inter-Knowledge

It is useful to distinguish between two different sources. It is not only important to know when the information is available, but also how and where it comes from. Figure 1.2 shows a simplified flow of a cognitive network. On the left side, there is knowledge from the node itself (*Intra-Knowledge*), and on the other side, the knowledge that has to be transmitted or measured from other nodes (*Inter-Knowledge*).

Intra-Knowledge (inside own node)
  A node itself can obtain a priori as well as a posteriori knowledge. It knows its configured gain, clipping threshold and the number of symbols to transmit (mapping scheme). Environment knowledge like system depth and ambient noise it can gain by itself, too. Further, this information and the knowledge from prior transmissions can even be stored and used for the next transmission attempt (posteriori knowledge). It is possible to get information without feedback from other

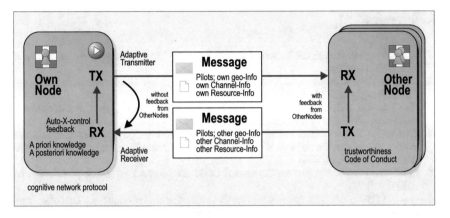

Fig. 1.2 Cognitive Network Protocol [27]

nodes. So, the echos arise from reflection could be used as a quality marker for the right source level. If for example an echo, detected by the transmitter itself, can decode the signal well, then the probability is higher that also the receiver can hear it well. On the other side, if the echo is too loud (clipped) or distorted it is presumably corrupted at the receiver. The code of conduct can, in the area of own knowledge, have a high impact, since all nodes have the same network protocol. If the *ownNode* knows what the communication participants can understand or how they behave on particular messages, it can adapt its behavior without sending unnecessary packets.

Inter-Knowledge (with support of other nodes)

The second source to get information are the *otherNodes* in the network. The main part of most studies on adaptive and cognitive networks focuses on this kind of knowledge. The necessary information about the channel state is exchanged via messages between the nodes or even while transmitting. The own Intra-Knowledge (like geographical location and channel-/resource-info) is distributed with predefined message packets to the other nodes. This can for example be the *Channel Impulse Response*. As the underwater channel is very time varying, this kind of information can very soon be outdated and useless. For this problem, pilots are interspersed to get the Doppler shift and delays within the transmission. The code of conduct is also useful as network-knowledge. Since all nodes have the same set of behavior rules, a feedback can be derived implicitly, like the forwarding rule in GUWMANET® [31], for example. If a transmitter sends a message and receives the forwarded packet after a certain time, it can be sure that its message has been heard.

The next chapters summarize the current works and state-of-the-art in cognitive underwater networks.

References

1. Richardson WJ, Greene CR Jr, Malme CI, Thomson DH (2013) Marine mammals and noise. Academic, New York
2. Imran A, Zoha A, Abu-Dayya A (2014) Challenges in 5g: how to empower son with big data for enabling 5g. IEEE Netw 28(6):27–33
3. Ha C-B, You Y-H, Song H-K (2019) Machine learning model for adaptive modulation of multi-stream in mimo-ofdm system. IEEE Access 7:5141–5152
4. Anjangi P, Chitre M (2018) Model-based data-driven learning algorithm for tuning an underwater acoustic link. In: 2018 fourth underwater communications and networking conference (UComms). IEEE, pp 1–5
5. COST 207 (1989) Management Committee. COST 207: digital land mobile radio communications
6. Nissen I (2008) Adaptive systems for mobile underwater communications with a p(oste)riori channel knowledge, first half. FWG report 59, Kiel
7. Mitola J (2000) Cognitive radio—an integrated agent architecture for software defined radio
8. Haykin S et al (2005) Cognitive radio: brain-empowered wireless communications. IEEE J Sel Areas Commun 23(2):201–220
9. Kolodzy PJ (2006) Interference temperature: a metric for dynamic spectrum utilization. Int J Netw Manag 16(2):103–113
10. Partan J, Kurose J, Levine BN (2007) A survey of practical issues in underwater networks. ACM SIGMOBILE Mob Comput Commun Rev 11(4):23–33
11. Shah CP, Tsimenidis CC, Sharif BS, Neasham JA (2010) Low complexity iterative receiver design for shallow water acoustic channels. EURASIP J Adv Signal Process 2010:12
12. Nissen I (2005) Pilot-based OFDM-systems for underwater communication applications. In: Proceedings conference on new concepts for harbour protection, littoral security and underwater acoustic communications, Istanbul, Turkey (TICA)
13. Carrascosa PC, Stojanovic M (2008) Adaptive MIMO detection of OFDM signals in an underwater acoustic channel. IEEE
14. Nordenvaad ML, Oberg T (2006) Iterative reception for acoustic underwater mimo communications. In: OCEANS 2006. IEEE, pp 1–6
15. Emre Y, Kandasamy V, Duman TM, Hursky P, Roy S (2008) Multiinput multioutput of dm for shallow-water uwa communications. J Acoust Soc Am 123(5):3891
16. Roy S, Duman TM, McDonald V, Proakis JG (2007) High-rate communication for underwater acoustic channels using multiple transmitters and space-time coding: receiver structures and experimental results. IEEE J Ocean Eng 32(3):663–688
17. Li B, Huang J, Zhou S, Ball K, Stojanovic M, Freitag L, Willett P (2009) Mimo-ofdm for high-rate underwater acoustic communications. IEEE J Ocean Eng 34(4):634–644
18. Mason S, Berger C, Zhou S, Ball K, Freitag L, Willett P (2009) An ofdm design for underwater acoustic channels with doppler spread. In: 2009 IEEE 13th digital signal processing workshop and 5th IEEE signal processing education workshop. IEEE, pp 138–143
19. Falk M, Bauch G, Nissen I (2020) On channel codes for short underwater messages. Multidisciplinary Digital Publishing Institute - Information
20. Stojanovic M (2006) Low complexity OFDM detector for underwater acoustic channels. IEEE
21. Leus G, Van Walree PA (2008) Multiband ofdm for covert acoustic communications. IEEE J Sel Areas Commun 26(9):1662–1673
22. Sharif BS, Neasham J, Hinton OR, Adams AE (2000) A computationally efficient doppler compensation system for underwater acoustic communications. IEEE J Ocean Eng 25(1):52–61
23. Li B, Zhou S, Stojanovic M, Freitag L, Willett P (2008) Multicarrier communication over underwater acoustic channels with nonuniform doppler shifts. IEEE J Ocean Eng 33(2):198–209
24. Nissen I (2009) Adaptive systems for mobile underwater communications with a p(oste)riori channel knowledge, second half, FWG report, 61, Kiel

25. Stojanovic M, Catipovic JA, Proakis JG (1994) Phase-coherent digital communications for underwater acoustic channels. IEEE J Ocean Eng 19(1):100–111
26. Sozer EM, Proakis JG, Blackmon F (2001) Iterative equalization and decoding techniques for shallow water acoustic channels. In: MTS/IEEE oceans 2001. An ocean odyssey. Conference proceedings (IEEE Cat. No. 01CH37295), vol 4. IEEE, pp 2201–2208
27. Nissen I, Kramer F, Thalheim B (2019) S-BPM. Inf Model Knowl Bases XXX 312:137
28. Endrjukaite T, Dudko A, Jaakkola H (2019) Information Modelling and Knowledge Bases XXX, vol 312. IOS Press, Amsterdam
29. Nissen I (2005) Appendix within the book of Peter C. Wille, Sound images of the oceans, pp 443–450
30. Ramakrishna S, Nissen I (2012) Next generation cognitive system approaches in the underwater communication area. Appl Ocean Res 38:136–141
31. Goetz M, Nissen I (2012) GUWMANET-multicast routing in underwater acoustic networks. In: 2012 military communications and information systems conference (MCC). IEEE, pp 27–42

Chapter 2
Adaptivity at the Physical Layer

**Henry Dol, Koen Blom, Paul van Walree, Roald Otnes, Håvard Austad,
Till Wiegand, and Dimitri Sotnik**

Link adaptation, or adaptive modulation and coding (AMC) (see Sect. 2.2), is a term used in wireless communications to denote the matching of the modulation, coding and other signal and protocol parameters to the conditions on the (radio) link, e.g., the path loss, the interference due to signals coming from other transmitters, the sensitivity of the receiver, the available transmit power margin, etc. [1]. AMC is about reconfiguring transmission characteristics and waveforms to exploit all opportunities, using model and/or duplex-feedback knowledge, shown in Fig. 2.1. For example, in radio communications, WiMAX uses a rate adaptation algorithm that adapts the modulation and coding scheme according to the quality of the (radio) channel [2].

For underwater acoustic communication, physical-layer (PHY) adaptivity at the transmitter (TX) side may involve in-mission optimizations of, for example:

- Modality (acoustic/optical/radio);
- Modulation (coherent/incoherent);
- Source level/transmit power;
- Frequency band (range vs. rate; availability);
- Data rate (vs. robustness: Quality-of-Service (QoS));
- Message length or size (nr. of blocks);
- Symbol constellation (e.g., BPSK/QPSK/8-PSK/16-QAM mappings);

H. Dol · K. Blom
Netherlands Organisation for Applied Scientific Research (TNO),
The Hague, Netherlands

P. van Walree · R. Otnes · H. Austad
Norwegian Defence Research Establishment (FFI), Kjeller, Norway

T. Wiegand
ATLAS ELEKTRONIK GmbH, Bremen, Germany

D. Sotnik
Fraunhofer-Institut für Kommunikation, Informationsverarbeitung
und Ergonomie (FKIE), Bonn, Germany

© The Author(s) 2020   13
D. Sotnik et al. (eds.), *Cognitive Underwater Acoustic Networking Techniques*,
SpringerBriefs in Electrical and Computer Engineering,
https://doi.org/10.1007/978-3-662-61658-1_2

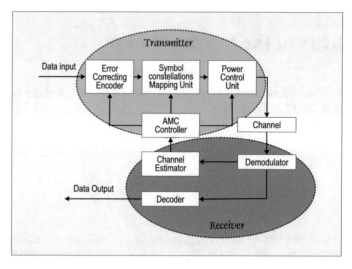

Fig. 2.1 Typical block diagram of a modem with Adaptive Modulation and Coding (AMC) [2]

- Pulse shape (roll-off factor; pulse shape function);
- Coding rate (Forward Error Correction (FEC));
- Training length (initial, periodic);
- Etc.

In addition, also the receiver (RX) characteristics can be made adaptive:

- Gain (via auto-gain control, or by number of hydrophone channels);
- Number of equalizer taps (i.e., filter lengths, e.g., based on observed time spread);
- Equalizer tap update algorithm (e.g., LMS/RLS);
- Phase-locked loop (PLL) settings, etc.

The modem settings can be optimized based on *user requirements* or *application settings* (APP, Fig. 2.2), e.g., the maximum source level, the number of bytes in the (unfragmented) message, the size of the Doppler filter bank (max. rel. speed), or the required Quality-of-Service (QoS). Settings can also be optimized based on the *environmental conditions* (ENV, Fig. 2.2), where we can distinguish between *physical descriptors*, such as input SNR (signal-to-noise ratio), ambient noise level, delay and Doppler spread (e.g., estimated via in-situ channel soundings), channel coherence time and time-of-flight (for ranging between network nodes), and *system descriptors*, such as output SNR, Doppler shift, estimated Bit Error Ratio (BER) and the degree of clipping.

In between APP and "ENV" are the network layer (NET) and the physical layer (PHY), including medium access control (MAC). The environmental 'measurements' are performed at the physical layer, whereas the application information comes in through the network layer. The network layer has the best overview of the network topology and can probably best decide on the required source level

Fig. 2.2 Interactions
between different layers for
adaptive underwater
communication

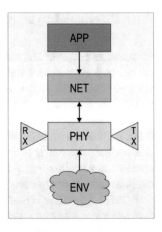

and hydrophone gain. To reduce the overhead, some local equalizer optimizations, such as (feedback/forward) filter lengths, tap update schemes and PLL settings, may probably best be decided about at the physical layer. It is expected that there will be different optimal receiver settings for different frequency bands. Other adaptions, that need to be synchronized between transmitter and receiver, should preferably also be managed by the network layer. These include for example the frequency band (LF/HF), modulation scheme, data rate (via number of redundant subbands, symbol constellation size, channel coding) and equalizer training sequences.

The adaptions can be communicated from TX to RX by separate messages (large capacity, latency penalty), by preamble coding (small capacity, message overhead), or perhaps even by classification of the received waveform (increased computational complexity).

Yet another level of adaptivity is the optimization of the, horizontal and/or vertical, position of a mobile node. An example of an AUV that searches for the best communication depth is given by [3]. The switching between LF and HF bands depending on the communication range (and SNR) is another example.

As a reading guide for the following sections, they aim to answer the following questions:

- What to adapt? (Sect. 2.1)
- Based on what information? (Sect. 2.2)
- How to communicate the adaptions between transmitter and receiver? (Sect. 2.3)

Finally, Sect. 2.4 addresses negotiation and switching mechanisms, optimization methods and decision criteria.

2.1 Physical-Layer Adjustments

Underwater acoustic channels are non-stationary on time scales relevant to typical network deployment times. There are many possible causes, such as changing weather conditions, tidal and diurnal cycles, and a variable range in the case of mobile nodes. Ambient noise in the ocean [4] is also non-stationary over longer deployments, as factors such as sea state and shipping density will vary with time. A physical-layer (PHY) solution with fixed parameter settings will be wasting energy when it operates in a regime where it is over-dimensioned, or result in disconnected nodes when it operates in a regime where it is under-dimensioned. For this reason, adaptive PHY methods have recently been proposed for underwater acoustic modems.

Popular adjustable signal parameters for transmitting are the source level and/or data rate, depending on whether the link quality is limited by SNR or delay-Doppler spread. The source level can be tuned in the SNR case, whereas the rate may help in both cases.

2.1.1 Adapting the Source Level

The source level of an acoustic modem is a measure of the emitted acoustic power, expressed in dB re $1\,\mu Pa^2 m^2$ [5]. Lowering the source level can be useful for conserving energy in the nodes and for reducing interference and medium occupancy, whereas an increase in source level will improve the quality of SNR-limited links. There have been several proposals for adaptive power control [6–9].

In [8], a transmission power control algorithm is implemented as an integral part of MAC and routing. When a node needs to transmit, it will send an Request To Send (RTS). The RTS contains some routing information, but important to the power control is the presence or absence of a Clear To Send (CTS). If no CTS from the receiving node is received at the transmitting node, it will increase its power level one step up (starting from step 0). It is unclear if the node stores its last successful power step, or if it starts from step 0 every transmission.

In [9], another power control algorithm is proposed. This is managed only by the Medium Access Control (MAC) layer, which is a hybrid of a channel reservation protocol and slotted MAC. Instead of RTS/CTS, the protocol uses RTS/CSI for channel reservation. When a node has a packet to transmit, it will send an RTS with maximum power, containing information about the TX power settings. The receiving node will calculate a channel gain based on the information in the RTS packet and the received power. The transmission power is adjusted according to the CSI packet and used when transmitting the data packet. RTS/CSI packets are sent as broadcast, so all overhearing nodes have the ability to build themselves a channel state matrix. For this matrix to be created, its assumed that the CSI packets also are transmitted with maximum power.

In many cases, it is most energy efficient to use as low a transmission power as possible (as long as you reach a neighbouring node) and compensate for this by using a multi-hop network. This is true from a selfish source node's perspective, but in an underwater sensor network most of the generated traffic is intended for a sink, in most cases a gateway solution. In a scenario where one or a few nodes are located very close to this sink, the most energy-efficient way for the *network* may not be to always use the lowest TX power reaching only one neighbour, as this can drain the node close to the sink [10].

2.1.2 Adapting the Data Rate

The term data rate is often synonymous with bit rate, whose basic definition is the number of bits that are conveyed per unit of time [11]. Acoustic modems and networks use packetized information with significant overhead, allowing multiple interpretations of the basic definition. From a user's point of view, a useful definition of bit rate is [12] "the amount of packetized information (available to higher protocol layers) expressed strictly as a number of bits each of which is is drawn from the set {0, 1}, divided by the total time required to transmit the packet."

The most popular ways to vary the data rate are through the symbol constellation size (alphabet) [13–19], spreading code length [13, 19–21] and channel coding redundancy [13, 16, 18, 22, 23].

In [16], Wan et al. suggest an AMC scheme with a finite set of OFDM modes and coding rates. The transmitting node will first send an RTS and the receiving node will reply with a CTS. The CTS will report the channel conditions which the transmitting node will use to determine what transmission mode will used. The metric used to switch between the different AMC modes is the Effective Signal-to-Noise Ratio (ESNR), which is calculated after channel estimation and is equivalent to the receiver output SNR.

In [22], Demirors et al. test a software-defined modem with a physical-layer adaptation that uses a chirp-based feedback link for switching and updating transmitter parameters. The evaluation criterion of the link is the Signal-plus-Interference-to-Noise Ratio (SINR). The parameters switched between are OFDM modulation order (constellation size) and error-correction code rate. In [14], the authors use received input SNR to determine the constellation size used in the next transmission.

The spreading code length in a Direct-Sequence Spread Spectrum (DSSS) system is used to adapt the data rate in [21]. The shortest spreading code length ($\widehat{=}$ highest data rate) that satisfies a BER constraint is chosen. This paper evaluates its proposed approach based on simulations only. Adapting the spreading code length is also proposed in [20].

In [19], Pelekanakis and Cazzanti propose to adapt both the signal constellation and the spreading code length. The potential is evaluated based on real-world physical-layer performance but the adaptation is simulated.

Tomasi et al. propose in [23] to adapt the coding redundancy based on channel conditions, using different adaptation algorithms. They evaluate the potential gains of their proposed methods based on KAM11 sea trial data.

Mani et al. propose in [18] to adapt the modulation order as well as the code rate, in a turbo-coded Phase Shift Keying (PSK) scheme also applying turbo equalization (with feedback loop from the decoder). They also propose to individually adapt these properties of multiple transmit streams in a Multiple Input Multiple Output (MIMO) scenario. As in other papers mentioned above, the potential of the proposed method is evaluated by simulations based on actual physical-layer performance from a real-world sea trial.

Many commercial modems provide a variety of modulation schemes and data rates, but implemented methods to adapt the data rate are rarely available off-the-shelf. In plenary discussions at the UComms 2018 conference, US participants stated that adaptive data rate techniques would be subject to strict export restrictions from the USA.

2.1.3 Other Physical Layer Adaptations

Other candidates for physical-layer adaptation are swapping between coherent and incoherent modulation schemes [20, 24, 25], redistribution of energy across OFDM carriers [15, 16, 26], and adjusting the number of transmitters in a MIMO setting [18].

Benson et al. [24] proposed already in 2000 to switch between coherent (PSK/ Quadrature Amplitude Modulation (QAM)) and incoherent (Frequency Shift Keying (FSK)) modulation schemes. They proposed to use the channel spread factor (delay spread multiplied by Doppler spread) to select the modulation scheme. If the spread factor was below 0.001, they would used a coherent scheme, and if it was above 0.001, they would use FSK.

In [25], Petroccia et al. propose to choose between three different modulation schemes, but with only one data rate per scheme: incoherent FSK at $\sim$2 b/s, coherent BPSK at $\sim$20 b/s, and a coherent trellis coded modulation at $\sim$100 b/s. The choice of modulation scheme is integrated with the routing protocol. The cost function takes energy consumption into account, considering that it takes more energy to transmit a given amount of data at a lower data rate when the transmit power is the same.

Demirors et al. [22] superficially mention the possibility of switching between DSSS and OFDM modulation, in addition to other topics mentioned in Sect. 2.1.2. In [20], Wu et al. describe efforts towards designing an adaptive modem which supports both FSK and DSSS modulation.

In [27], Shankar and Chitre mention that the UNET-II modem of Acoustic Research Lab in Singapore can switch between incoherent and coherent modulation schemes, but the paper focuses on tuning (adaptation) schemes.

Adaptive power allocation between OFDM subcarriers is thoroughly treated by Pottier et al. in [26], in a scenario where nodes interfering with one another are considered. The nodes use an iterative water-filling approach in a non-cooperative game-theoretic approach, with the goal of reaching a Nash equilibrium.

Radosevic et al. propose in [15] to adapt the power levels across OFDM sub-carriers combined with the modulation scheme. For each subcarrier, a power level and modulation scheme would be chosen suitable to the predicted conditions in that subband.

A different form of adaptivity can be realized by optimizing receiver parameters for a fixed modulation format. For instance, the study by Benson et al. [24] uses a wideband channel probe to estimate delay spread and Doppler spread, which are subsequently used to determine the number of equalizer taps and channel tracking parameters, respectively. This approach does not involve overhead to communicate changes between nodes, as the receiver tries to figure out optimal parameters to demodulate a fixed modulation format. On the other hand, there is significant overhead in the form of the channel probe. It is expected that such parameter tuning can yield a moderate gain, and that changing the data rate, or swapping between coherent and incoherent schemes, will have a larger impact. User experiences with FRSS tell that one should not expect miracles from tuning the equalizer length, but there is one example on record where the strength of the PLL had to be increased to get more packets through.

Another option at the receiver is to vary the Least-Mean-Squares (LMS) step-size parameter or the Recursive-Least-Squares (RLS) forgetting factor for a fixed equalizer length, or even to implement a hybrid solution with RLS for the training symbols and LMS for the payload symbols [28]. This is not necessarily done to deal with a changing environment, e.g., the motivation may be just to find a balance between equalizer convergence and computational complexity. However, the potential for adaptivity is there.

2.2 Input Parameters for Physical-Layer Adaptivity

For underwater modems to operate in (extremely) variable conditions, Adaptive Modulation and Coding (AMC) and flexible higher-level protocol parameters are crucial for reliable and bandwidth-efficient communications. This section focuses on gathering meaningful information to adapt the parameters of the modulation and coding scheme.

What is meant by 'meaningful information' for AMC? To prevent an over-dimensioned or under-dimensioned modulation and coding scheme (*i*) flexibility to optimize parameters based on the user, application and platform context are necessary, and (*ii*) parameter optimization based on environmental awareness needs to be explored.

For now, assume that all information necessary for AMC is gathered by an adaptivity module (AMC control unit) within the network layer. User, application and platform information *descends the stack* from the application layer (APP) towards the adaptivity module in the network layer (NET), as shown in Fig. 2.3. The user,

Fig. 2.3 Direction of
information flow for
Adaptive Modulation and
Coding (Adaptive
Modulation and
Coding (AMC))

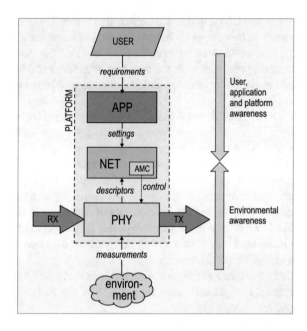

application and platform information relevant for AMC is discussed in Sect. 2.2.1.
Information related to the environmental awareness *ascends the stack* from the physical layer (PHY) to the adaptivity module in the network layer. Section 2.2.2 elaborates
on information related to the environmental awareness of the physical layer.

   To enable adaptation at the transmitter, the receiver needs to inform the transmitter. Section 2.3 elaborates on the realization of this feedback loop. During the
March 2018 National Science Foundation (NSF) Workshop on Underwater Wireless
Communications and Networking, the design and demonstration of feedback-based
acoustic systems was identified as an important research topic [29].

2.2.1 Relevant User, Application and Platform Inputs for Adaptation

The user, application and platform inputs for adaptation are dependent on the use
case, not on real-time awareness. Since these inputs are not as rapidly changing as
the environment conditions, it is expected that the time scale of these adaptations is
much longer.

Maximum source level

   Based on the battery status, the application or user may restrict the maximum
   source level in order to extend the battery life.

Message size

The size of the message that should be transmitted unfragmented is a typical requirement of the user/scenario/application at hand. This may vary from a single block (i.e., minimum message size) for command & control messages to many blocks for images. The total number of (user) bits per message could be a multiple of 128 bit, such as in GUWAL/GUWMANET. Ideally, to reduce overhead, messages are not fragmented and multiple 128-bit PHY blocks are transmitted in a single message.

Maximum platform velocity

Movement of the source, scatterers and the receiver will result in Doppler frequency shifts of the source signal. In underwater communications, the acoustic bandwidth is (often) in the order of its carrier frequency. Therefore, Doppler cannot be compensated by mere frequency shifts. Doppler on wideband pulses manifests itself as a temporal scaling of the transmitted waveform.

For detection of communication waveforms, receivers often use a filter bank of Doppler-shifted preamble replicas. The range of Doppler velocities for such a filter bank are application dependent. E.g., a network of stationary bottom nodes does not require the same Doppler range as a dynamic platform such as an Autonomous Underwater Vehicle (AUV). Under-dimensioning the range of the Doppler Filter bank can result in communication waveforms not being detected, whereas over-dimensioning results in wasting computational resources at the receiver, or in having a coarser Doppler resolution than necessary, which can decrease performance.

Quality-of-Service (QoS)

To maximize bandwidth utilization underwater, the Quality-of-Service (QoS) of the data to be transmitted should be taken in to account. QoS is a user/application requirement and it affects both the physical-layer and network-layer configuration. E.g., certain applications are delay-sensitive, while other ones are more concerned with average transmission rate. Service requirements are often expressed using the metrics (i) bandwidth, (ii) delay, (iii) jitter and (iv) loss rate. A more fine-grained overview of QoS metrics is given in Fig. 2.4. QoS adaptability will require information to be used across layers [30], e.g., the physical layer must be informed of the desired QoS of the payload to be transmitted. The tolerance for bit errors is dependent on the type of data being transmitted. Recent coding techniques allow digital voice communication with an bit-error ratio of $\leq 1 \cdot 10^{-2}$ [32]. For image data holds that, using the right techniques to reduce and conceal the effects of transmission errors, a bit-error ratio around $1 \cdot 10^{-4}$ is still acceptable [33]. For data communication in mobile radio networks, independent of the type of information, $\leq 1 \cdot 10^{-6}$ [34] can be tolerated. In these networks, retransmission schemes can be used to guarantee bit-error-free message transfers.

In some literature, mostly in the field of wireless broadcasting, the term Quasi-Error Free (QEF) is used. Depending on the application, the exact QEF bit-error ratio differs. E.g., in modulation schemes used by Digital Video Broadcasting (DVB) standards, QEF transmission is defined as a bit-error ratio of $\leq 1 \cdot 10^{-10}$ [35].

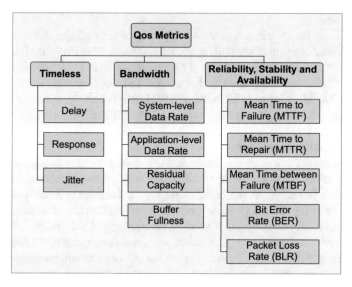

Fig. 2.4 Taxonomy of QoS metrics [31]

An example of an open-water experiment that focused on exploiting QoS for underwater communication is shown in Fig. 2.5. During this fjord trial, a ship node broadcasted Multi-Stream Frequency-Repetition Spread Spectrum (MSFRSS) transmissions containing on-board webcam snapshots and GPS) NMEA strings. These payloads acted as the low-QoS and high-QoS data, respectively. Details on this trial and the MSFRSS modulation can be found in [36].

2.2.2 Adaptation Based on Environmental Awareness

Environmental awareness of the receiver can be based on two types of input: *physical and system-based descriptors*. A selection of relevant physical and system-based descriptors is discussed in the next sections.

2.2.2.1 Physical Descriptors of the Link Quality

Independent of the exact modulation and its implementation, physical descriptors of the link quality can (for each reception) be calculated and used as input for AMC.

Instead of adaptation based on the current environmental awareness, Radosevic et al. [15] propose to adapt based on predicted conditions. Their adaptive schemes were tested on recorded test channels and real-time during sea trials. The link quality is

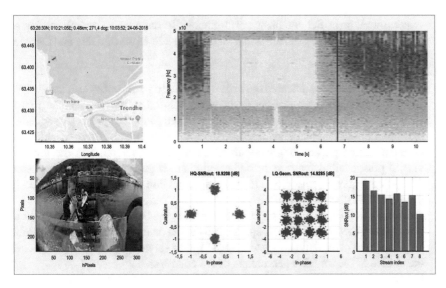

Fig. 2.5 Real-time reception of GPS and image data in QPSK/16-QAM mode

predicted by estimating the channel state parameters for the next transmission with
an RLS algorithm [7, 15]. An enhanced system throughput is demonstrated.

Input SNR

The common definition for Signal-to-Noise Ratio (SNR), or input SNR, is the
ratio of average signal power to average noise power, where the noise power is
measured in the band of the signal. To obtain unambiguous measurements by
different modems, ideally the noise power should be determined using the same
bandpass filter as used for the transmit signal (e.g., for Frequency-Repetition
Spread Spectrum (FRSS), this is a root-raised cosine response). Additionally, the
averaging intervals and the definition of 'frequency band' (e.g., -3 dB or -6 dB
points) should be agreed upon [37]. In [38], Heidemann et al. discuss the length
of the aforementioned averaging intervals. Their viewpoint is that large-scale
variations should mainly influence power control, whereas small-scale variations
should mainly influence the choice of modulation and coding.

Most underwater communication systems are not noise-limited, but are limited
by the inability to fully compensate for the time-variant response of the acoustic channel, resulting in self-interference. In systems with self-interference, the
receiver *output* SINR is more practical than the input SNR. For now, the self-
interference in the output SINR is treated as noise and we will refer to it as
output SNR. It (inversely) scales with the size of the constellation clouds and is

thus a measure of the equalizer performance. To determine the link quality of time-variant responses, channels can be expressed in terms of delay and Doppler spread.

Doppler spread

Modem signals can be smeared out in the frequency domain because of transceiver motion, sound speed fluctuations in the water column, or moving scatterers such as surface gravity waves. This smearing is called Doppler spread. An accurate measurement of Doppler spread is hard to obtain at the receiver from a wideband communication packet, unless significant overhead is tolerated in the form of known pilot symbols/carriers or dedicated channel probe signals. Transmission of a probe sequence, in addition to the communication waveform, would severely decrease the effective data rate.

Benson et al. [24] use a wideband probe signal at the beginning of each packet to measure the delay and Doppler spread. No details on their probe signal (or probe sequence) are given. In their system, the Doppler spread measurement influences the choice of the tracking parameters of the equalizer.

Delay spread

Acoustic modem signals are normally received over multiple paths with different propagation delays. The resulting delay spread is often in the millisecond regime, depending on the environment and the employed frequency band, but it can also be much longer. The power distribution in delay can be sparse or dense. The matched-filter output of the preamble detector (or the matched-filter output of a sounding sequence) can be used to gain insight in the delay spread of the channel. In Benson et al. [24], the delay spread, which is determined by means of the wideband probing signal, influences the length of the channel equalizer.

Channel spread factor

The channel spread factor is the product of the Doppler spread and the delay spread. It has been discussed as a means for adaptation in Sect. 2.1.3.

Ambient noise level

The difference in ambient noise levels between the low-frequency and high-frequency bands can be used to determine whether a low-frequency or high-frequency transmission is preferred. E.g., if a ship is passing by, a low-frequency transmission could be switched to the high-frequency band (if the receiver is nearby) to make sure that the transmission can be correctly received.

Channel coherence time

The coherence time T_{coh} is the time interval over which the channel coherence drops by some specified amount. Its inverse value gives the rate at which (coherent) communication receivers need to update their channel estimates. The phase of a signal propagated through the channel is predictable over the coherence time, but not beyond. I.e., if the phase is known at $t = t_0$ it is known up to $t = t_0 + T_{coh}$. This picture is of course not black and white. In reality the decorrelation is gradual. A scenario that is often encountered in acoustic channels is that part of the signal power is received over stable paths, and another part over fluctuating paths. This may require a definition of coherence time tailored to the application.

Fig. 2.6 The detection margin of SNR in a communication system [41]

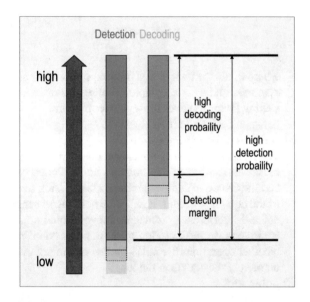

Inter-node distance

Based on time-of-flight measurements, the distance between nodes can be esti-mated. Knowledge of the distance between nodes can be helpful for band switch-ing and the scheduling of transmissions. Distance estimation is discussed in chapter "Delay/Disruption Tolerant Networking".

2.2.2.2 System-Based Descriptors of the Link Quality

System-based descriptors of the link quality are strongly coupled to the modem implementation and configuration. An example of a system-based descriptor is the output SNR. A selection of system-based descriptors is discussed in this section.

Output SNR

Output SNR is a measure of the fidelity of the receiver output. Acoustic modems usually operate at an output SNR that is well below the input SNR, unless the modulation scheme has a large processing (spread-spectrum) gain. The reason is that receivers are not able to harvest all the signal power. Paths with excess prop-agation delays or excess Doppler shifts are out of reach, and the power received over such paths acts as self-interference.

Different modulation schemes may require different computation methods, and even for a given scheme there may exist multiple definitions of output SNR [37]. Minimum mean-square error (MMSE) receivers, like the FRSS multiband equalizer, attempt to maximize the output SNR by minimizing the Mean Square Error (MSE) between the transmitted source symbols and the received output symbols. This gives rise to a bias [39]. FRSS uses the following definition to calculate the output SNR:

$$\text{SNR}_{\text{out}} = 10 \log_{10} \left[\frac{\mathbb{E}\{|z_n|^2\}}{\mathbb{E}\{|\hat{z}_n/\gamma - z_n|^2\}} \right] , \qquad (2.1)$$

where z_n are the transmitted training symbols, $\hat{z}_n$ are the corresponding equalized symbols, and γ is a complex scaling factor calculated as $\gamma = \mathbb{E}\{\hat{z}_n/z_n\}$. This scaling factor removes the equalizer bias and restores the theoretical relationship between SNR and bit error ratio [37] .

Signal and noise contributions

The equalizer output symbols contain a contribution of two (pseudo) random processes, the uniformly-distributed symbol stream (with residual self-interference) and the Gaussian additive ambient noise. Using second- or higher-order moments of the equalizer output, it might be possible to separate the contribution of residual self-interference, known as convolutional noise, and the contribution of the additive noise [40]. Notion of these noise contributions could (maybe) indicate whether operations are noise-limited or limited by the channel conditions, based on merely the equalizer output.

Doppler shift

In most receivers, an estimate of the average Doppler frequency shift of the received waveform is available from the detector. As mentioned in Sect. 2.2.1, the detector uses this frequency shift to compensate for the temporal scaling of the waveform. Due to movement of the transmitter, receiver and the sea surface, the Doppler shift varies with time.

Number of detections versus correct demodulations

The number of detected preambles versus correct demodulations is an indirect measure of the channel and ambient noise conditions. If signals are detected, but not correctly demodulated, then the modulation needs to be more robust to compensate for the channel conditions (Fig. 2.6). A Cyclic Redundancy Check (CRC) could be used to determine whether a message is correctly demodulated or corrupted. Some authors even estimate the Bit Error Ratio (BER), e.g., using special pilot messages [27].

Viterbi decoder metric

In a Nokia patent [42, expired], the channel quality is determined by looking at the Viterbi decoder metric for the maximum likelihood path. This Euclidean distance metric is filtered using a moving average filter to smooth out short-term variations. According to the patent, the filtered metric is a reliable indicator for the channel quality independent of the modulation scheme. Viterbi decoding is also applied by the FRSS modem.

Degree of clipping
 The receiver should be able to detect that the signal coming from the Analog-to-Digital Converter (ADC) is clipped. Feedback that the modem is clipping can be used as a trigger to reduce the amplification (gain) by the receiver and/or to trigger the transmission of feedback messages to the transmitter to request for a lower source level (Sect. 2.1.1).

2.3 Communication of Adaptions

There are basically three ways to communicate changes to the link settings between nodes:

1. Explicitly by separate configuration feedback messages;
2. Explicitly as part of the regular messages, e.g., by preamble encoding;
3. Implicitly by waveform detection and/or classification.

Going from 1 to 3, robustness decreases (especially from 2 to 3), parameter space decreases (especially from 1 to 2), and latency decreases (especially from 1 to 2).

2.3.1 Configuration Messages

The basic approach that was originally proposed [43, 44] to communicate changes is to transmit dedicated feedback messages. This is by far the most flexible approach as the whole (feedback) message's payload size may be available for communication of (arbitrary) changes. On the other hand, this approach also involves most increase of latency due to the additional messages required, although they may not be needed very often. Furthermore, a vulnerability is that, when a configuration message is not received (correctly), all following messages cannot be received (correctly) until the next configuration message is received (correctly).

Most publications dealing with adaptive (underwater) communications apply explicit configuration via a feedback link, see, e.g., [21, 22, 26, 45–47] for recent literature, where [26, 45, 46] report the use of a feedback link for power (and rate) control. Using Software-Defined Radio (SDR) technology, [22] applies a chirp-based feedback link to adapt modulation constellation and channel coding rate, or to switch between OFDM and DSSS modulation. Note that, when a different modulation is applied for the feedback link, this change needs to be detected by the receiver (Sect. 2.3.3). To minimize the overhead of the feedback messages, some authors apply compressed sensing or related methods [48, 49]. References [21, 47] apply (rate/coding) adaptation structures at the physical-layer frame or slot level, indicating that the feedback frame should be communicated within the channel's coherence time.

2.3.2 Preamble Encoding

In order to decrease the latency and vulnerability to lost configuration messages, changes can also be communicated inside the messages to which the changes apply. This should then be encoded in the message preamble since demodulation of the following message payload will require knowledge of the changes. When using the existing detection preamble, there will not be much capacity for changes to be communicated, as the preamble overhead should be minimized and the autocorrelation properties needed for detection (and synchronization) should not be impaired too much. Therefore, only a few settings that may change relatively often should be preamble encoded. A setting to which this may apply is, for example, the message size (number of blocks). More settings may be communicated by grouping the possible changes into a limited number of profiles, such that only the profile number needs to be encoded in the preamble.

Instead of encoding the profile number in the detection preamble, which may require running several detectors in parallel, and/or which may reduce the autocorrelation gain, it may also be part of an additional info block inserted between the detection preamble and the message. In fact, this is the common approach in terrestrial communication (IEEE 802.11, "Wireless Local Area Networks"). This info block should have a fixed and robust modulation and may also contain other important (cross-layer) information (e.g., network address). Of course, the addition of an info block will reduce the effective data rate, which is more of concern underwater than for radio. Since this addition applies to every message, and not only to the (ad-hoc) configuration messages, the total overhead may easily become larger than for the feedback approach.

2.3.3 Waveform Detection and Classification

No increase of message overhead and latency occurs when the receiver can automatically detect and/or classify changes in the received waveform (self-awareness/blind parameter estimation). Good examples is the automatic detection of an incoming message in one of the available frequency bands (LF and HF in [50]), provided that detection is performed in all bands or the detection of short bursts with different signal lengths based on the synchronization estimation [51]. Also, the differences between the waveforms for different modulations (e.g., JANUS vs. FRSS), the number of (redundant) frequency bands (e.g., of FRSS), the symbol constellation (e.g., BPSK, QPSK, 8-PSK, 16-QAM, etc.) [52, 53] and the symbol rate may be distinguished by classification techniques, see for example Fig. 2.7. These changes often lead to a change of the data rate (blind-rate detection [47]). Overviews of likelihood- and feature-based (often using higher-moment statistics) classification methods relevant for adaptive communication are given by [53–56]. The price to be paid is the

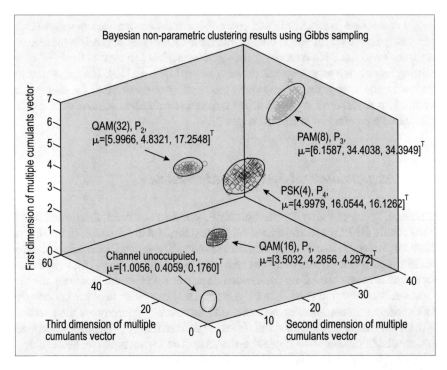

Fig. 2.7 Pattern discovery and clustering results of multiple cumulants vectors, for classifying the constellation size [56]

additional detection/classification effort by the receiver and sensitivity to the (input) SNR.

In fact, computational overhead may increase significantly for the classification approach, and the experience (of WTD 71) is that blind communication parameter estimation based on modeling (a-priori knowledge) often leads to wrong settings (too many unknowns) and thus inefficient channel usage.

2.4 Adaptive Parameter Control

A major challenge in Parameter Control (PC)/negotiation (e.g., of coding rate, modulation scheme, transmit power, frequency band, etc.) for digital underwater acoustic communications is to find an appropriate Performance Indicator (PI) (e.g., received SNR, estimated BER/PER, channel conditions, etc.) [16] and a suitable PC method to reach predefined criteria for a Communication Link (CL). Possible CL criteria are for example the link data rate, BER/PER and/or target SNR (e.g., low target SNR is important for covert communications). The trade-off between these criteria needs to be quantified [19] and the parameter control needs to be optimized. Therefore, a correlation between PIs and controllable transmission parameters needs to be investigated. Unfortunately, PIs like BER and PER highly depend on the channel conditions and there is no closed-form solution available to predict the BER of an existing modulation scheme in the underwater environment [19]. Hence, even if the environmental conditions are known by the receiver or transmitter, for example by feedback, it is still challenging to select a specific modulation scheme or to modify transmission parameters to reach a target BER.

2.4.1 Negotiation and Switching Mechanisms

In general, two opposite system designs for PC can be considered. The first approach is to execute the PC based on measured PIs and forward the decision (transmission mode or parameter) to other communication nodes. The second approach is to transmit PIs (e.g., sparse channel impulse response) in a feedback link [15]. In that case, each node executes the PC by itself based on received PIs. In the literature, the first approach is more often used. The PC method itself can differ between cooperative nodes within a communication network, without destroying interoperability, as long as the overall design and system interfaces are defined (e.g., PC interfaces, feedback link, available transmission modes/parameters). The PC method can be developed depending on the processing capability of each node.

In the literature, there are a few PC methods investigated, which can be roughly divided into two categories:

1. Switching to a predefined transmission mode/profile (e.g., combination of modulation scheme and code rate, different frequency bands) depending on a fixed PI threshold (in most cases SNR) [14, 16, 18, 24, 48, 57].
2. Transmission parameter control or mode selection to optimize predefined CL criteria (in most cases data rate and/or BER).

The first category has little computational complexity, but also limited adaption capabilities in a changing environment. More sophisticated approaches aim to optimize predefined CL criteria by satisfying different PIs. These approaches are clustered in the second category and will be considered in the following section.

2.4.2 Optimization Methods and Decision Criteria

Generally, the PC optimization problem aims to maximize or minimize the CL criteria in compliance with further constrains [17, 21, 22, 46]. As an example, Fig. 2.8 by Proakis [58] shows the relationship between bandwidth efficiency and power efficiency for different modulation schemes at a symbol error probability of 10^{-5}. It can be observed that the bit-rate to bandwidth ratio increases with the number of signal points, at the expense of an increased SNR per bit [58].

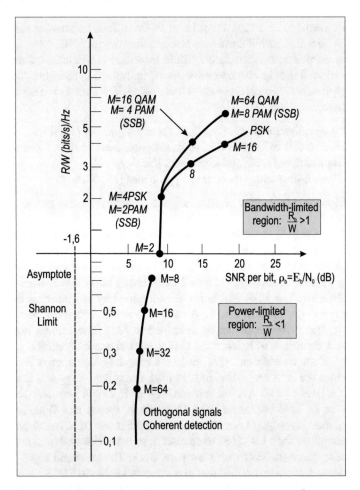

Fig. 2.8 Comparison by Proakis [58] of several modulation methods at 10^{-5} symbol error probability

A PC optimization problem is formulated in [17]:

$$\max_{M,C} R(M, C), \qquad (2.2)$$

satisfying

$$BER_{th} \geq BER(SINR, M, C). \qquad (2.3)$$

As shown in (2.2) and (2.3), [17] aims to maximize the data rate R as a function of the coding rate C and modulation scheme M to satisfy a specified BER threshold. BER is a function of M, C and SINR. The authors in [16] propose to use the effective SNR (ESNR) instead, which is calculated after channel estimation. According to

[16], ESNR seems to be a robust metric for PC in different channel conditions. As indicated in Sect. 2.1, ESNR corresponds to receiver output SNR.

In many publications, there is only little information available on the specific implementation of the algorithm to solve the PC optimization problem. A number of different approaches found in literature is listed below. The approaches are sometimes used in a hybrid fashion.

1. Classical optimization algorithms (e.g., greedy algorithm [15, 46]).
2. Data-driven/empirical approaches (e.g., decision trees [59], boosted trees [19], bandit algorithms [27, 60], reinforcement learning [61]).
3. Model-based/information-theoretical approaches [16, 60, 61].

The following paragraphs contain a brief summary of the previously listed approaches.

Radosevic et al.[15] propose two different PC schemes. Scheme 1 aims to adaptively allocate the modulation scheme M while distributing the transmit power P uniformly among the subcarriers of an OFDM transmission and maintaining a target BER using a greedy algorithm. Scheme 2 additionally allocates the transmit power of each subcarrier. The BER thresholds are computed by an empirical function of M, coding rate C and channel gain G. A comparable approach is considered in [46].

A purely data driven approach is described in [27]. The authors use a multi-armed bandit algorithm to balance the exploration versus exploitation problem in an OFDM system. In addition to [27, 60] uses a model-based approach in a hybrid fashion to tune the communication link. In [59], the authors suggest a decision tree to predict the BER of a specific modulation scheme in a single-carrier transmission, depending on channel parameters like SNR, delay spread and Doppler. In [19], the same author investigates boosted trees to predict the BER based on different channel parameters for a low-SNR transmission scheme. Boosted trees combine the learning technique AdaBoost with regression trees. The baseband signal in [19] is simultaneously modulated on different carriers comparable to FRSS.

The work in [61] investigates a model-based approach. The controllable transmission parameters are P, M and C. The approach is based on recursive channel model parameter estimation, based on channel measurements. The optimization problem includes continuous and discrete variables and is solved by Monte Carlo planning and aims to minimize a long-term system cost.

2.5 CSI Estimation, Prediction and Adjustment

The Performance Indicator is needed to decide on the fitting parameter sets. Mostly SNR is used with Feedback Channels and Channel Impulse Response, in [48, 62–64]. Because, as in most other communication techniques regardless of underwater communications, it is necessary to know the channel-state information (CSI). This requires a procedure of measuring and estimation. Often, a differential detection is performed in traditional radio networks. As its implementation significantly degrades frame-error rate versus signal-to-noise ratio, it is not practical [64]. Alternatively, classical Pilot transmission has to be sent prior to data communication and is therefore not only a wasteful signal, but for the time-varying underwater environment, it could be outdated within the transmission. A Pilot transmission during the signal, shown in [13, 65], could provide side information using an orthogonal matching-pursuit algorithm. As all these techniques and processes require additional space in the packet transmissions or extra signals, they are not very efficient to find suitable parameters.

In order to find meaningful parameters, it is necessary to look into the future. As is mentioned in Sect. 1.4, it is a smart idea to look also at own knowledge like data from prior transmissions or self-inflicted echos based on the signal-to-reverbration-ratio to optimize the parameters. For that, it is indispensable to track information. In current studies, there are different approaches of tracking and predicting the CSI. Everyone pursues other objectives and tracks according to different criteria, i.e. energy saving [63, 64, 66] or robustness [48, 67]. Haykin suggests transmitting a training sequence (2 to 4 known symbols) prior to data transmissions [64]. He advises to track the time-varying autoregressive coefficients and the corresponding noise level. In a non-Gaussian environment, a *particle filter* can be used to avoid disturbance variables [68]. The prediction can then be performed by a *higher-order Markovian model*. The transmission power is then decided with a game theory model (Nash-Equilibrium) or a so-called *Iterative Water-Filling* [64] to achieve a global optimum.

In [69], a machine learning (ML) model is presented for tracking and prediction in a 5G Network, with a variable number of available modulation schemes and antennas. Since the rule-based techniques will be reaching their limits of maintenance with increasing numbers of distinct network nodes, a learning algorithm is proposed. The relationship between channel conditions and optimal transmission mode can be learned by supervised training prior to communication. The algorithm then can classify more than one optimal mode at the same SNR, to achieve more adaptability than a simple rule-based method. But since there are often only limited hardware capacities for underwater sensors and ML is very ambitious, a more capable solution for UWSN was shown in [66]. This ML-based Routing protocol has low overhead, since it only uses the information from direct neighbours, like averaged residual energy. The proposed unsupervised Q-Learning algorithm (based on Reinforcement Learning) uses this residual energy as target to choose the best path in the network to save energy.

A Model-based Data-driven Learning Algorithm is shown in a recent publication [67]. It utilizes both the information gained from CSI (Delay spread, coherence time, Doppler spread, ...) and techniques from data-driven algorithms. By computing the probability of packet success, the key parameters in OFDM (number of sub-carriers and cyclic prefix) could be chosen by a confidence Upper bound algorithm to maximize the data rate. If the delay spread and the coherence time is not given during the message transmission, the algorithm also allows default values, suitable to the environment.

Qiao et al. in [65] follow another interesting approach. In addition to the CSI gained from the pilot transmission, mentioned above, they propose an adaptive downlink

OFDMA system with low overhead and limited feedback. A data fitting method for channel reconstruction is described, which shows better performance than the traditional group quantization method (tested in the South China sea). To indicate an *outdated CSI*, they define a long-term statistical mean and a *per-subcarrier channel temporal correlation* (PTSCTC). These are used to decide between an average CSI and the currently observed CSI.

An estimation technique based on compressed sensing, where the CSI feedback is only from sparse channel parameters, is shown in [48]. "The objective of the channel prediction is to estimate the future values of the fading coefficient ahead". They predict this values every several symbols with a linear channel prediction algorithm. Without prior information of the maximum Doppler shift or scattering, the estimation is based only on amplitude, delay of each path, and channel impulse response. An autoregressive model is proposed which computes the minimum mean square error estimate of a future fading coefficient. Depending on this linear algorithm, a fixed threshold adaption is performed and the modulation mode (BPSK, QPSK, 8PSK, 16PSK with two-dimensional gray mapping) was chosen. Simulations and tests in an experimental pool at the Xiamen University (China) show that significant improvements can be obtained for BER and throughput compared with a fixed QPSK modulation scheme.

2.6   Adaptive MAC

An adaptive network consists roughly of estimation and prediction of the Channel State Information (CSI). In the last chapter, was shown the current techniques and processes which unite these two, to get a good decision base. To achieve the best adjustment, it is necessary to know how the channel behaves. But we also have adaptivity at higher layers in the protocol stack. Flooding-based protocols are inherently adaptive to changing network topologies, and both the GUWMANET and DFlood protocols from EDA RACUN project have forwarding rules which adapt to network conditions. It might also be possible to adapt protocol parameters to traffic load experienced in the network. For example, it has been shown [70] that the number of allowed re-transmissions should be reduced as traffic load increases.

One can envision adaptivity at the medium access control (MAC) layer.

With respect to medium access control (MAC), it is known that different approaches perform best in different scenarios. E.g., handshake-based methods (e.g., request/clear-to-send, RTS/CTS) outperforms direct random access (e.g., ALOHA with carrier sensing) if packet sizes are large compared to propagation delay, or in dense and busy networks, while direct random access is better when packet sizes are short, or in sparse networks with low traffic [71, 72]. Similarly, time-division multiple access (TDMA) is inefficient unless packet sizes are large compared to maximum propagation delay in the network.

It has been proposed to dynamically adapt the choice of handshaking or direct random access to network conditions (e.g., packet size) [71, 73]. It has also been proposed to adapt handshaking delay parameters to measured propagation delays [74], and to adapt the choice of MAC protocol to capabilities of peer nodes in a heterogeneous network [75]. Apart from these efforts, not much can be found in open literature about adaptive MAC protocols.

The MAC protocol used mostly will be ALOHA with carrier sensing, which in RACUN was found to be most suitable for typical underwater scenarios with short packets. This may be altered if it is found that other choices would be better in adapting to the scenarios at hand. If other MAC protocols than ALOHA are used (e.g., protocols that adapt to measured propagation delays, or TDMA protocols), it may be considered to improve their performance by employing high-accuracy chip-scale atomic clocks (CSACs) in the nodes.

2.7 Summary of Adaptivity Parameters and Methods

The tables below summarize the communication parameters, input parameters and ways to communicate adaptions, as discussed in this chapter.

What to adapt? (Section 2.1)

• transmitter side
 	
    – source level / transmit power (across carriers)
    – data rate
 	
       · symbol constellation size (mapping)
       · spreading code length
       · channel coding redundancy (coding rate)
 	
    – modulation type (in/coherent)
    – message size (unfragmented, nr. of blocks)
    – frequency band
    – number of redundant subbands (FRSS)
    – pulse shape (e.g. used in Filtered Multi-Tone, FMT)
    – training length (initial, periodic)
 	
• receiver side
 	
    – detection threshold
    – receiver gain
    – filter lengths (of equalizer)
    – tap update algorithm (of equalizer)
    – phase-locked loop settings (of equalizer)

Based on what information? (Section 2.2)

- user/scenario requirements, application settings
    - max. source level (based on battery status)
    - message size (unfragmented, nr. of blocks)
    - max. (relative) platform speed (nr. of Doppler channels)
    - Quality-of-Service (QoS: data rate vs. robustness)
- environmental conditions
    - physical descriptors
        · input SNR
        · ambient noise level
        · delay spread
        · Doppler spread
        · channel coherence time
        · inter-node distance
    - system descriptors
        · output SNR (convolutional vs. ambient noise)
        · Doppler shift
        · BER estimation
        · degree of clipping

How to communicate adaptions? (Sections 2.3, 2.4)

- feedback messages (Inter-Knowledge)
    - decision taken and communicated (local decision)
    - relevant input parameters communicated (remote decision)
        · switch to predefined mode (profile)
        · in-mission optimization
- preamble encoding
    - profile encoded in detection/synchronization preamble
    - profile encoded in info block following detection/synchronization preamble
- classification of received waveform (Intra-Knowledge)

References

1. Link adaptation, wikipedia page. https://en.wikipedia.org/wiki/Link_adaptation. Accessed 22 March 2019
2. Abdallah S, Martin M, Chadi S (2009) Broadband access networks. Springer, Berlin
3. Schmidt H, Schneider T (2016) Acoustic communication and navigation in the new arctic, a model case for environmental adaptation. In: 2016 IEEE third underwater communications and networking conference (UComms), pp 1–4. IEEE
4. Dahl PH, Miller JH, Cato DH, Andrew RK (2007) Underwater ambient noise. Acoust Today 3(1):23–33
5. Underwater acoustics — Terminology. ISO 18405:2017(E), International Organization for Standardization, Geneva, CH, 2017
6. Rice JA, McDonald VK (1999) Adaptive modulation for undersea acoustic telemetry. Sea Technol 40(5):29–36
7. Qarabaqi P, Stojanovic M (2011) Adaptive power control for underwater acoustic communications. In: Proceedings OCEANS 2011 IEEE - Spain, pp 1–7
8. Jornet JM, Stojanovic M, Zorzi M (2010) On joint frequency and power allocation in a cross-layer protocol for underwater acoustic networks. IEEE J Ocean Eng 35(4):936–947
9. Yishan S, Zhu Y, Mo H, Cui J-H, Jin Z (2015) A joint power control and rate adaptation MAC protocol for underwater sensor networks. Ad Hoc Netw 26:36–49
10. Bouabdallah F, Zidi C, Boutaba R (2017) Joint routing and energy management in underwater acoustic sensor networks. IEEE Trans Netw Ser Manag 14(2):456–471
11. Bit rate (2018) https://en.wikipedia.org/wiki/Bit_rate. Accessed 14 March 2018
12. van Walree P, Green D, Otnes R (2018) Ambiguities in underwater acoustic communications terminology and measurement procedures. In: 2018 fourth underwater communications and networking conference (UComms), pp 1–4
13. Nissen I (2005) Pilot-based OFDM-systems for underwater communication applications. In: Proceedings conference on new concepts for harbour protection, littoral security and underwater acoustic communications, Istanbul, Turkey (TICA)
14. Tomasi B, Toni L, Casari P, Rossi L, Zorzi M (2010) Performance study of variable-rate modulation for underwater communications based on experimental data. In: Proceedings of the OCEANS 2010 MTS/IEEE SEATTLE, pp 1–8
15. Radosevic A, Ahmed R, Duman TM, Proakis JG, Stojanovic M (2014) Adaptive OFDM modulation for underwater acoustic communications: design considerations and experimental results. IEEE J Ocean Eng 39(2):357–370
16. Wan L, Zhou H, Xu X, Huang Y, Zhou S, Shi Z, Cui J (2015) Adaptive modulation and coding for underwater acoustic OFDM. IEEE J Ocean Eng 40(2):327–336
17. Demirors E, Sklivanitis G, Santagati GE, Melodia T, Batalama SN (2014) Design of a software-defined underwater acoustic modem with real-time physical layer adaptation capabilities. In: Proceedings of the international conference on underwater networks & systems, p 25. ACM
18. Mani S, Duman TM, Hursky P (2008) Adaptive coding-modulation for shallow-water UWA communications. J Acoust Soc Amer 123(5):3749
19. Pelekanakis K, Cazzanti L (2018) On adaptive modulation for low SNR underwater acoustic communications. In: OCEANS 2018, SC, USA, November 2018
20. Lingjuan W, Trezzo J, Mirza D, Roberts P, Jaffe J, Wang Y, Kastner R (2012) Designing an adaptive acoustic modem for underwater sensor networks. IEEE Embed Syst Lett 4(1):1–4
21. Gannon A, Balakrishnan S, Sklivanitis G, Pados DA, Batalama SN (2018) Short data record filtering for adaptive underwater acoustic communications. In: 2018 IEEE 10th Sensor array and multichannel signal processing workshop (SAM), pp 316–320, July 2018
22. Demirors E, Sklivanitis G, Santagati GE, Melodia T, Batalama SN (2018) A high-rate software-defined underwater acoustic modem with real-time adaptation capabilities. IEEE Access 6:18602–18615
23. Tomasi B, Munaretto D, Preisig JC, Zorzi M (2015) Redundancy allocation in time-varying channels with long propagation delays. Ad Hoc Netw 34:31–41

24. Benson A, Proakis J, Stojanovic M (2000) Towards robust adaptive acoustic communications. In: OCEANS 2000 MTS/IEEE conference and exhibition, vol 2, pp 1243–1249. IEEE
25. Petroccia R, Pelekanakis K, Alves J, Fioravanti S, Blouin S, Pecknold S (2018) An adaptive cross-layer routing protocol for underwater acoustic networks. In: 2018 fourth underwater communications and networking conference (UComms), pp 1–5. IEEE
26. Pottier A, Socheleau F-X, Laot C (2017) Robust noncooperative spectrum sharing game in underwater acoustic interference channels. IEEE J Ocean Eng 42(4):1019–1034
27. Shankar S, Chitre M (2013) Tuning an underwater communication link. In: 2013 MTS/IEEE OCEANS - Bergen, pp 1–9
28. Freitag L, Johnson M, Stojanovic M (1997) Efficient equalizer update algorithms for acoustic communication channels of varying complexity. In: Oceans '97. MTS/IEEE conference proceedings, vol 1, pp 580–585
29. Song A, Stojanovic M, Chitre M (2019) Underwater acoustic communications: where we stand and what is next? (editorial). IEEE J Ocean Eng 44(1):1–6
30. Berezdivin R, Breinig R, Topp R (2002) Next-generation wireless communications concepts and technologies. IEEE Commun Mag 40(3):108–116
31. Malik A, Qadir J, Ahmad B, Alvin Yau K-L, Ullah U (2015) QoS in IEEE 802.11-based wireless networks: a contemporary review. J Netw Comput Appl 55:24–46
32. Cheruku DR (2010) Satellite communication. IK International Pvt Ltd
33. Abdat M, Al Kachouh Z, Bellanger MG (1998) Transmission error detection and concealment in JPEG images. Signal Process Image Commun 13(1):45–64
34. Minoru E (2005) Next generation mobile systems: 3G and beyond. Wiley, New York
35. Tozer EPJ (2012) Broadcast engineer's reference book. Taylor & Francis, Abingdon-on-Thames
36. Blom KCH, Dol HS (2018) QoS-enabled underwater acoustic communications. In: OCEANS 2018 MTS/IEEE charleston, pp 1–7. IEEE
37. van Walree P (2013) On the definition of receiver output snr and the probability of bit error. In: 2013 MTS/IEEE OCEANS-Bergen, pp 1–9. IEEE
38. Heidemann J, Stojanovic M, Zorzi M (2012) Underwater sensor networks: applications, advances and challenges. Philos Trans R Soc A: Math Phys Eng Sci 370(1958):158–175
39. Cioffi JM, Dudevoir GP, Eyuboglu MV, Forney GD (1995) MMSE decision-feedback equalizers and coding–Part I: equalization results. IEEE Trans Commun 43(10):2582–2594
40. Matzner R, Englberger F (1994) An SNR estimation algorithm using fourth-order moments. In: Proceedings of 1994 IEEE international symposium on information theory, p 119. IEEE
41. Berning F (2019) Thoughts regarding adaptive modulation profile switching for SALSA. SALSA presentation
42. System and method for measuring channel quality information. US Patent 6108374 (filed in 1997, granted in 2000)
43. Hayes JF (1968) Adaptive feedback communications. IEEE Trans Commun Technol 16(1):29–34
44. Chua S-G, Goldsmith AJ (1997) Variable-rate variable-power MQAM for fading channels. IEEE Trans Comm 45(10):1218–1230
45. Ahmed R, Stojanovic M (2017) Joint power and rate control for packet coding over fading channels. IEEE J Ocean Eng 42(3):697–710
46. Sadeghi M, Elamassie M, Uysal M (2017) Adaptive OFDM-based acoustic underwater transmission: system design and experimental verification. In: 2017 IEEE international black sea conference on communications and networking (BlackSeaCom), pp 1–5
47. Nanda S, Balachandran K, Kumar S (2000) Adaptation techniques in wireless packet data services. IEEE Commun Mag 38(1):54–64
48. Kuai X, Sun H, Qi J, Cheng E, Xiao-ka X, Guo Y, Chen Y (2014) CSI feedback-based CS for underwater acoustic adaptive modulation OFDM system with channel prediction. China Ocean Eng 28(3):391
49. Gang Q, Liu L, Ma L, Yin Y (2019) Adaptive downlink ofdma system with low-overhead and limited feedback in time-varying underwater acoustic channel. IEEE Access 7:12729–12741

50. Dol H, Colin M, van Walree P, Otnes R (2018) Field experiments with a dual-frequency-band underwater acoustic network. In: 2018 fourth underwater communications and networking conference (UComms), pp 1–5. IEEE
51. Nissen I (2015) Burst communication-a solution for the underwater information management. Hydroacoustics 18:113–126
52. Xing Z, Zhou J, Ye J, Yan J, Zou L, Wan Q (2014) Automatic modulation recognition of PSK signals using nonuniform compressive samples based on high order statistics. In: 2014 IEEE international conference on communication problem-solving, pp 611–614. IEEE
53. Thakur PS, Madan S, Madan M (2015) Trends in automatic modulation classification for advanced data communication networks. Int J Adv Res Comput Eng Technol (IJARCET) 4:496–507
54. Dobre OA, Abdi A, Bar-Ness Y, Wei S (2007) Survey of automatic modulation classification techniques: classical approaches and new trends. IET Commun 1(2):137–156
55. Hazza A, Shoaib M, Alshebeili SA, Fahad A (2013) An overview of feature-based methods for digital modulation classification. In: 2013 1st international conference on communications, signal processing, and their applications (ICCSPA), pp 1–6. IEEE
56. Gao F, Zhang K (2015) Enhanced multi-parameter cognitive architecture for future wireless communications. IEEE Commun Mag 53(7):86–92
57. Kochanska I (2013) Considerations of adaptive digital communications in underwater acoustic channel. Hydroacoustics J
58. Proakis MSJG (2002) Communication systems engineering, 2nd edn. Pearson, London
59. Pelekanakis K, Cazzanti L, Zappa G, Alves J (2016) Decision tree-based adaptive modulation for underwater acoustic communications. In: 2016 IEEE third underwater communications and networking conference (UComms), pp 1–5
60. Anjangi P, Chitre M (2018) Model-based data-driven learning algorithm for tuning an underwater acoustic link. In: 2018 fourth underwater communications and networking conference (UComms), pp 1–5
61. Wang C, Wang Z, Sun W, Fuhrmann DR (2018) Reinforcement learning-based adaptive transmission in time-varying underwater acoustic channels. IEEE Access 6:2541–2558
62. Carrascosa PC, Stojanovic M (2008) Adaptive MIMO detection of OFDM signals in an underwater acoustic channel. IEEE
63. Patricia Ceballos Carrascosa and Milica Stojanovic (2010) Adaptive channel estimation and data detection for underwater acoustic mimo-ofdm systems. IEEE J Ocean Eng 35(3):635–646
64. Haykin S et al (2005) Cognitive radio: brain-empowered wireless communications. IEEE J Sel Areas Commun 23(2):201–220
65. Qiao G, Liu L, Ma L, Yin Y (2019) Adaptive downlink ofdma system with low-overhead and limited feedback in time-varying underwater acoustic channel. IEEE Access
66. Tiansi H, Fei Y (2010) Qelar: a machine-learning-based adaptive routing protocol for energy-efficient and lifetime-extended underwater sensor networks. IEEE Trans Mob Comput 9(6):796–809
67. Anjangi P, Chitre M (2018) Model-based data-driven learning algorithm for tuning an underwater acoustic link. In: 2018 fourth underwater communications and networking conference (UComms), pp 1–5. IEEE
68. Haykin S, Huber K, Chen Z (2004) Bayesian sequential state estimation for mimo wireless communications. Proc IEEE 92(3):439–454
69. Ha C-B, You Y-H, Song H-K (2019) Machine learning model for adaptive modulation of multi-stream in mimo-ofdm system. IEEE Access 7:5141–5152
70. Otnes R, van Walree P, Buen H, Song H (2015) Underwater acoustic network simulation with lookup tables from physical-layer replay. IEEE J Ocean Eng 40(4):822–840
71. Xie P, Cui J-H (2006) Exploring random access and handshaking techniques in large-scale underwater wireless acoustic sensor networks. In: OCEANS 2006, pp 1–6. IEEE
72. Otnes R, Asterjadhi A, Casari P, Goetz M, Husøy T, Nissen I, Rimstad K, van Walree P, Zorzi M (2012) Underwater acoustic networking techniques. Springer Science & Business Media, Berlin

73. Zhang W, Qin Z, Xin J, Wang L, Zhu M, Sun L, Shu L (2014) Upmac: a localized load-adaptive mac protocol for underwater acoustic networks. In: 2014 23rd international conference on computer communication and networks (ICCCN), pp 1–8. IEEE
74. Guo X, Frater MR, Ryan MJ (2009) Design of a propagation-delay-tolerant mac protocol for underwater acoustic sensor networks. IEEE J Ocean Eng 34(2):170–180
75. Shahabudeen S, Chitre M, Potter J, Motani M (2009) Multi-mode adaptive mac protocol suite and standardization proposal for heterogeneous underwater acoustic networks. In: Proceeding underwater acoustic measurement, pp 1–8

Chapter 3
Distance Estimation

Ivor Nissen

Positioning and localization aspects for mobile underwater vehicles in the underwater network are fundamental but critical parts. In the important standard scenario a node has to estimate a distance between its single-hop-neighborhood. Figure 3.1 displays this process in a network of communication nodes. An AUV is entering the network and is passing bottom node A in the third barrier. The AUV should transmit a amount of data on a higher frequency band. Afterwards, the AUV leaves the neighborhood of barrier three with node A and continues its track to node B in the fourth barrier. Here the AUV stops at node B for data muling (Sect. 4.4). An efficient Data Muling process needs a decision, when the Data mule is close enough to the sensor node to start the muling process. Also the methods for Multi Topology Routing needs a decision, when the subject has to switch from low frequency band to the higher band.

Some subjects like bottom nodes are equipped with nuclear clocks, others not. In the area of un-synchronized neighbors, alternative strategies are possible. To solve it, there exists a number of different approaches and solutions, and each of them have advantages and disadvantages. The simplest one is to ask the other nodes for the position and than to calculate the Euclidean distance to the own position. This needs the transmission of position messages, costs therefore energy. The question is now, what are possible active and passive tactics to determinate distances for a cognitive networking localization strategy.

Acoustic localization is a fundamental function in the underwater environment, since it plays a critical role in many applications—especially in underwater networks. A rough search in the Google Scholar references database (https://scholar.google.de/) with the keywords 'underwater network distance measurement' generates more than 101,000 references; with the keywords 'self localization underwater networks' more than 25,100 references. One of the first proposed methods was the so-called 'ping-pong' method, first used for underwater navigation in Norwegian Fjords to estimate distances between the ship and the rocks. The time diagram for this method is shown in Fig. 3.2, left, for the "MARK-SNAP" method:

I. Nissen
Wehrtechnische Dienststelle für Schiffe und Marinewaffen, Maritime Technologie und Forschung (WTD 71), Eckernförde, Germany

© The Author(s) 2020
D. Sotnik et al. (eds.), *Cognitive Underwater Acoustic Networking Techniques*,
SpringerBriefs in Electrical and Computer Engineering,
https://doi.org/10.1007/978-3-662-61658-1_3

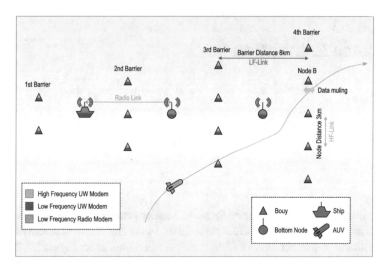

Fig. 3.1 The topology for this scenario with moving AUV that estimates distance to bottom sensor nodes

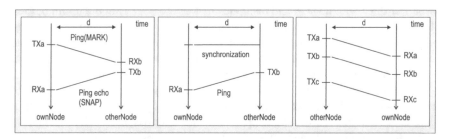

Fig. 3.2 MARK-SNAP method (left, Submarine manual MXP1-Table 5–9) for underwater telephones, if messages include time stamps, the sync error could be estimated via nonlinear equations (right)

General range-based algorithms typically consist of the common three steps:

- range measurement to the neighbor subjects (relative link distance),
- location estimation (e.g., multi-lateration for global position),
- calibration of parameters.

The ranging process is based on the knowledge of the other nodes. For the exchange of position information, network protocols are not needed. The so-called IN-RES protocols (without network headers) can perform the first steps with less traffic load. The INitiator transmits its request, the RESponders answer by using the checksum of the request parcel, to bind the cooperation tasks together [1] into a conversation. But inside of a cognitive network protocol a large diversity of realization possibilities are existing.

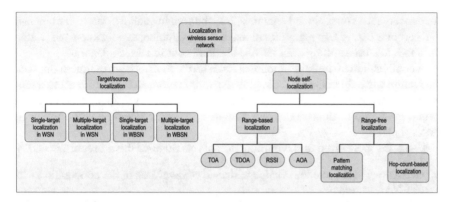

Fig. 3.3 Taxonomy of localization methods independed of the media, [2]. W(B)SN is the Wireless (Binary) Sensor Network

In this section different approaches, described in the literature, are presented:

In the review article [2] from Cheng, two localization method groups, independent of the media (above and under water)—mainly for the terrestrial radio-based systems, are being classified: target/source localization and node self-localization methods. Target localization is based on energy. The node self-localization using

- localization in non-line-of-sight,
- node selection criteria for localization in energy-constrained networks,
- scheduling the sensor node to optimize the trade-off between localization performance and energy consumption
- cooperative node localization and
- localization algorithm in heterogeneous networks

implies a complex taxonomy of methods, displayed in Fig. 3.3.

Evaluation criteria for localization in a wireless sensor network are described using different common metrics like the average localization error, root mean square error, and geometric mean error. The localization errors are inevitable in the estimations.

For the underwater domain localization, baseline systems [long baseline system (LBL), short baseline system (SBL), ultra-short and super-short baseline systems (USBL and SSBL)] are used with transmissions of short SONAR pings in the multi-path, half-duplex nature of the wet environment by multiple synchronous nodes in different frequency bands. These transmissions are no communication signals—the conventional 'ping', a short-time CW pulse or sweep, contains no payload information. With the measured time and an adapted constant and fixed sound speed value (e.g., equal 1500 m/s for different places/depths), relative distances are estimated, based on, e.g., time difference of arrival approaches (TDoA). All devices are typically time synchronized. A variety of effects in the underwater column complicate this determination. The sound speed is varying from approx. 1407–1570 m/s over space (depths) and time [3], and it is partly unknown to the receiver. The same is valid for the transmission losses caused by distances over the 'line of sight'-ray (especially for convergence zones, sound channels, over bottom in shallow waters, archipelago and harbor areas, …). In many situations, it is not sufficient to only know the relative distances to a transmitting unit, instead absolute coordinates are desired.

A combination of digital communication with the transmission of time stamps and navigation can help in the absolute global determination, and therefore act as a kind of an "underwater GPS". The global position process needs distance measurements with a minimum of three nodes and a multi-lateration estimation as the following step.

Using the view given in the introduction of this State-of-the-art report, Sect. 1.4,

- Inter-Knowledge—the knowledge collected of *ownNode* in the cooperation with *otherNodes*—and the
- Intra-Knowledge—the self-collected knowledge without cooperation with other subjects

generates the structure of this section of distance estimation methods.

3.1   Inter-Knowledge

In the article [4], five main challenges in underwater localization are described, based on

Node Deployment   Deployment of sensor and reference nodes is difficult and costly in the deep sea environment.

Node Mobility   Underwater nodes are being drifted away by underwater current and other activities. The speed of the current is variable over time and difficult to predict. If a node is moved during the localization process, its position estimation goes wrong.

Change in Signal Strength   The strength of the acoustic signal gets affected by many factors like Doppler shift, multi-path propagation, attenuation and external noise.

Time Synchronization   Many localization schemes assume that nodes are synchronized. Time synchronization is difficult to achieve in an underwater scenario due to long propagation delay and variable sound speed. Radio signals cannot propagate underwater, GNSS services likewise GPS or Galileo are not available under the surface.

Variation in Sound Speed   Localization schemes assume constant sound speed, but in reality it is variable and depends on temperature, pressure and salinity. Variations in any one of these factors alter the sound speed. It may introduce an error in the distance estimation, and the accuracy of the scheme may decrease.

Acoustic Channel Characteristics   The speed of sound in water is five orders of magnitude lower than of radio waves in air. Therefore, the propagation delay is higher. Bandwidth is limited and it depends on transmission range and frequency. The data rate are low compared to radio and Bit Error Ratio (BER) is higher.

Table 3.1 Table for the comparison of localization schemes, [4]

Scheme	Rangebased/ Rangefree	Range Measurement Using	Time Synchronization Required	Silent Positioning	Node Mobility Considered	Iterative Localization Used
DNR	range based	TOA	yes	yes	yes	no
PL	range based	TOA	yes	yes	yes	yes
LSL	range based	TOA	yes	yes	yes	yes
AFLA	range based	TOA	yes	no	yes	no
LSLS	range based	TDOA	no	yes	no	yes
USP	range based	TOA	yes	yes	no	yes
3DUL	range based	two way message exchange	no	no	yes	yes
UPS	range based	TDOA	no	yes	no	no
WPS	range based	TDOA	no	yes	no	no
UDB	range free	N/A	no	yes	no	no
LDB	range free	N/A	no	yes	no	no
ALS	range free	N/A	no	yes	no	no

Two groups of localization methods are compared in this article from Beniwal [4]:

Range-based localization schemes are presented using range or bearing information for the position estimation. These may use Time of Arrival (ToA), Time Difference of Arrival (TDoA) or Received Signal Strength Indicator (RSSI) for the distance estimation process. Angle-of-Arrival (AoA) is missing in this paper.

The members of the second group are Range-free schemes which do not use range or bearing information. These are simple techniques and provide coarse-grained location estimation for underwater nodes. An overview with a comparison is included in Table 3.1 displaying 12 range-based, range-free, synchronization-required and iterative approaches.

Chen [5] propose a hierarchical localization scheme to address the challenging problems with limited bandwidth, the severely impaired channel and the cost of underwater equipment. In the paper, simulation results and additional analysis were presented.

In the article from Chandrasekh [6], range-based schemes (Infrastructure-based, Distributed Positioning, using Mobile Beacons/Anchors, without Anchor/Reference Points) and range-free schemes (Hopcount based, Centroid, Area-based Localization, Area Localization, Approximate Point In Triangle) are classified (Table 3.2).

The authors conclude "localization for terrestrial sensor networks has been studied in great detail. However, the problem of localization in underwater sensor networks poses a new set of challenges because of the acoustic transmission medium. [...] Many of the localization schemes discussed here are shown to work in simulation, and their performance needs to be evaluated in underwater systems. Finally, localization is discussed in the application domain of UWSNs in offshore engineering."

The survey Han et al. [7] divides the set of approaches in the underwater domain into three categories: Stationary localization algorithms, mobile localization algorithms and hybrid localization algorithms; the classification is shown in Fig. 3.4.

Table 3.2 Comparison of underwater localization methods, [6]

Schemes	Range based or Range Free	Accuracy	Distributed or Centralized	Placement of anchor nodes	% of anchor nodes	Additional Comments
Infrastructure bases positioning systems	Range based, ToA, TDoA	Accurate: 1 to 10m for 3km x 4km Area. Accuracy depends on area size	Distributed	At the corners of a square grid	Small	Requires placement of anchor nodes on sea-bed
Distributed positioning	Range based, ToA, TDoA	Not Accurate: 0,5° (Radio Range) to 1° (Radio Range)	Distributed	Distributed randomly	High (5% to 20% of nodes)	Requires placement of anchor nodes on sea-bed
Mobile Beacons	Range based, ToA	Accurate: < 1m, for schallow water of < 500m (Sonardyne)	Distributed	Only one anchor	Low	The mobile beacon should be a ship equipped with GPS, or an AUV/ROV whose location is known
DV-Hop	Range Free	Not Accurate: 0,5° (Radio Range) to 1° (Radio Range)	Distributed	At the corners of a square grid	Low	Simple to implement
Centroid based localization	Range Free	Not Accurate: 0,5° (Radio Range) to 1° (Radio Range)	Distributed	In a grid structure	High (High for a good performance)	Simple to implement, but requires placement of anchors in a square mesh
ALS	Range Free	Not Accurate: 0,5° (Radio Range) to 1° (Radio Range)	Centralized	At the corners of a square grid	Low	Anchor nodes must be able to cover area in consideration. Simple to implement
APIT	Range Free	Not Accurate: 0,5° (Radio Range) to 1° (Radio Range)	Distributed or Centralized	Randomly distributed	High (High for a good performance)	Anchor nodes must be able to cover area in consideration. Simple to implement
Fingerprinting, Signal Processing based schemes	Range based, RSSI	Accurate, but only good for small aereas	Centralized	No anchor nodes	N/A	Very difficult to implement in the underwater domain because of training phase

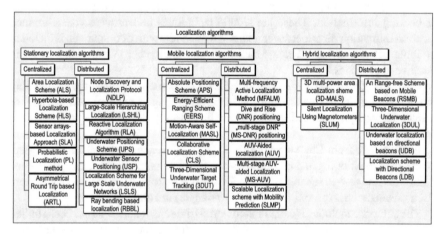

Fig. 3.4 Classification of localization methods based into three categories: stationary localization, mobile localization and hybrid localization algorithms, [7]

The authors conclude with "current localization algorithms are based on a safe and credible environment. However, in real applications, UWSNs are always deployed in complex and unsafe environments. Thus, secure localization and position verification algorithms are needed." And they continue with the statement: "Develop an efficient localization mechanism in which multiple anchor nodes dynamically collaborate with each other to localize unknown nodes. Especially needed is the research on how anchor nodes should collaborate to localize unknown nodes when the number of anchor nodes is not enough."

Ahmed and Salleh [8] focus in their survey on centralized and distributed localization schemes. Both of the schemes are further divided into estimation-based and prediction-based schemes. This survey focuses on the weaknesses of every scheme, analyzing and describing different parameters like synchronization, communication, range method and anchor types used by schemes. Furthermore, the survey points out the common open issues of the researchers in the field of localization.

The article [9] introduces applications like target tracking or disaster prevention, where sensed data is meaningless without location information. They propose a new 3D centralized localization scheme for mobile underwater wireless sensor networks, named Reverse Localization Scheme or RLS in short. RLS is an event-driven localization method triggered by detector sensors for launching the localization process as a good starting point for restricted flooding network protocols. RLS is suitable for surveillance applications that require very fast reactions to events and could report the location of the occurrence. Major contributions of this method lies in reducing the numbers of message exchanged for localization, saving the energy and decreasing the average localization response time. The document includes proposals for the protocol design with details and concrete algorithms (Fig. 3.5).

In the paper Cheng et al. [10], the authors present a silent positioning scheme termed UPS for underwater acoustic sensor networks. UPS relies on the time difference of arrivals locally measured at a sensor to detect range differences from the sensor to four anchor nodes. These range differences are averaged over multiple beacon intervals before they are combined to estimate the 3D sensor location through tri-lateration. UPS requires no time synchronization and provides location privacy at underwater vehicles/sensors whose locations need to be determined. The authors wrote "UPS is superior to existing systems in many aspects, such as lack of synchronization and low computation overhead. To evaluate the performance of UPS [...]. Our scheme is simple and effective." This approach has similarities to the Kieler approach (see [11, 12]).

The article from Tan [13] conducted a wide survey of recently proposed localization schemes specifically designed for UWSNs. The authors identified several of the challenges that need to be overcome for underwater localization schemes to be fast and accurate, have low communication costs, provide wide coverage and be feasible. In addition to classifying the schemes under

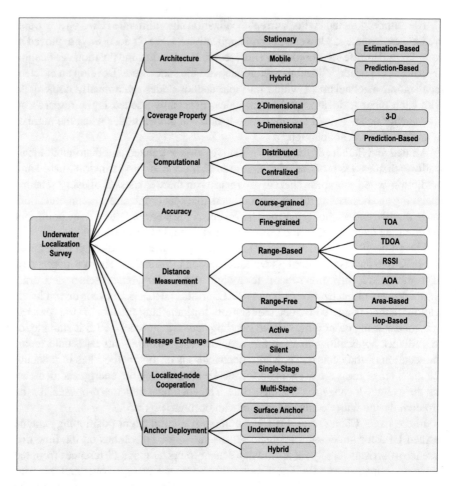

Fig. 3.5 A wider taxonomy tree for underwater localization methods, [9]

 (i) range-based, the authors also further classify it as

   • single versus multi-stage and
   • static versus mobile references.

 (ii) range-free and
(iii) finger-printing based schemes,

In Sect. 6 of this article open problems and research challenges are pointed out.

While challenges associated with the reference node deployment, time synchronization, and asymmetric power consumption in acoustic modems have been addressed to some extent in the proposed schemes. In the authors view, the following challenges should be, but have not been, fully addressed:

Sound Speed Variation   Most range-based localization techniques assume a constant speed of sound underwater.

Inter-node Time Synchronization   Localization schemes that rely on silent positioning to minimize communication overhead assume that nodes are time-synchronized.

Node Mobility Model   Node mobility due to water currents, which presents one of the greatest challenges for underwater localization, has only been accounted for up to various degrees.

Impact of Medium Access Control   Another important challenge that has not been fully addressed is Medium Access Control (MAC) to resolve contention, particularly in multi-stage localization schemes for dense UWSNs.

Impact of Channel Structure   The underwater acoustic channel is a frequency-selective time-varying channel. Since the localization process requires range measurement, using either ToA, TDoA or RSSI techniques, the structure of the channel may affect the accuracy of the localization process.

Performance Evaluation   The proposed range-based and range-free schemes have only been evaluated analytically or via numerical simulations. For a fair evaluation of the schemes, they should be evaluated using available simulators for underwater acoustic networks (e.g., Harris and Zorzi 2007) under a common simulation scenario.

These six points resemble the main challenges in underwater localization given in the Sect. 1.4 Inter-Knowledge.

In Peleato and Stojaconic [14], additional input for the design of MANET tactics are being presented. The authors propose under the name of Distance-Aware Collision Avoidance Protocol (DACAP) a non-synchronized protocol that allows a node to use different hand-shake lengths for different receivers so as to minimize the average hand-shake duration. Both authors propose two tactics to minimize the resulting inefficiencies in case a node overhears an RTS or a CTS from a hand-shake going on within its transmission range:

1. Nodes continue listening to the channel while in the back-off state, and return to the idle state if they detect the end of the current transmission (data packet or acknowledgment).
2. If a node receives an RTS from one of the two nodes involved in the current transmission, it replies with a CTS and exits the back-off state.

3.2 Intra-Knowledge

Given is a burst communication with the information of the transmitter position (like a lighthouse beacon) in combination with the delay of the first singular echo of the burst. With this knowledge and the multi-lateration calculation, a global position and the mean sound speed can be estimated, based purely on time measurements and the resample factor of the Doppler effected communication. Due to the short transmission time of bursts, it is even possible to decode echoes, for example reflections from the surface or sea floor.

OwnNode is a mobile and not cooperating subject in the water column without global position knowledge, in a depth of h, the water depths is h_B. *OwnNode* has to build up own knowledge. The Doppler resampling factor is given by

$$\frac{f_{RX}}{f_{TX}} = \frac{T_{TX}}{T_{RX}} = \frac{\bar{v} - |s_R| cos \delta_R}{\bar{v} - |s_T| cos \delta_T}$$

with the symbol time T_{TX} at the transmitter and the receiver T_{RX}, the unknown median sound speed in the water column $\bar{v}$ and the subject speeds s_R, s_T with their angles. In the case of $s_T = 0$ e.g. for bottom nodes (GUWAL message type = 01) this reduce to the equation:

$$\frac{f_{RX}}{f_{TX}} = \frac{T_{TX}}{T_{RX}} = 1 - \frac{|s_R| cos \delta_R}{\bar{v}} = 1 - \frac{1}{\bar{v}} \left\langle \begin{pmatrix} S_R^x \\ S_R^y \end{pmatrix} \middle| \frac{\begin{pmatrix} X0 - x \\ Y0 - y \end{pmatrix}}{\left| \begin{pmatrix} X0 - x \\ Y0 - y \end{pmatrix} \right|} \right\rangle ,$$

if two *otherNodes* exists and we can assume a constant median sound speed, bottom nodes with a fixed and given position or gateway buoy with GPS knowledge $X0$, $Y0$ and $X1$, $Y1$. Each node can collect the messages of the other nodes of the neighborhood. If sound speed values, latitude, longitude and other values are included in the messages, this information should be stored. *OwnNode* can estimate the distance to all known nodes in this manner (Kiel approach).

$$\bar{v} \left(\frac{T_{TX}}{T_{RX}} - 1 \right) = \left\langle \begin{pmatrix} S_R^x \\ S_R^y \end{pmatrix} \middle| \frac{\begin{pmatrix} x - X0 \\ y - Y0 \end{pmatrix}}{\left| \begin{pmatrix} x - X0 \\ y - Y0 \end{pmatrix} \right|} \right\rangle$$

Following the thesis of Dubrovinskaya Elizaveta [11] let L be the line of sight (LOS) distance, and R be the traveled distance of the reflected beam. At the left part (V-scenario) in Fig. 3.6 with both equations $\{d^2 + h^2 = L^2, (2h_B - h)^2 + d^2 = R^2\}$ and the substitution $\{L = \bar{v} * t_L, R = \bar{v} * (\delta_T + t_L)\}$ this leads to

$$H = h : \left\{ \bar{v} = \sqrt{\frac{4h_B(h_B - H)}{\delta_T(\delta_T + 2t_L)}}, d = \sqrt{t_L^2 \bar{v}^2 - H^2} \right\}$$

At the right part (A-scenario) in Fig. 3.6 with both equations $\{d^2 + (h_B - h)^2 = L^2, (h_B + h)^2 + d^2 = R^2\}$ and the substitution $\{L = \bar{v} * t_L, R = \bar{v} * (\delta_T + t_L)\}$ this leads to

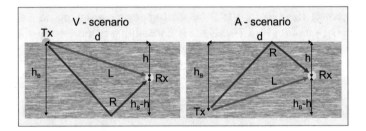

Fig. 3.6 Estimation of the distance via LOS and first echo, Kiel approach

$$H = h_B - h : \left\{ \bar{v} = \sqrt{\frac{4h_B(h_B - H)}{\delta_T(\delta_T + 2t_L)}}, d = \sqrt{t_L^2 \bar{v}^2 - H^2} \right\}$$

If the nodes are synchronized, it is easy to measure t_L (time for LOS sound ray to travel the distance L) and $\delta\tau_p = t_R - t_L$ (time difference between arrivals of LOS, sound ray and reflected sound ray). Having knowledge about full depth h_B and mobile node depth h, the horizontal distance and sound speed c_{sound} can be calculated (see Fig. 3.6). It is assumed that sound speed is constant, hence sound rays are straight. The value $\delta\tau_p$ could be estimated in the modem.

In case the nodes are not synchronized, the synchronization error can be estimated. Given are the last three TX transmissions from one unsynchronized node (*other-Node*, TX_a, TX_b, TX_c) including a time stamp. *OwnNode* receives the messages and decode them (RX_a, RX_b, RX_c). A simple equation system with three unknowns and the synchronization offset ψ:

$$RX_a = TX_a + \delta_a + \psi \quad RX_b = TX_b + \delta_b + \psi \quad RX_c = TX_c + \delta_c + \psi$$

and in the form of the travel times δ_a, δ_b and δ_c:

$$\delta_a = RX_a - TX_a - \psi \quad \delta_b = RX_b - TX_b - \psi \quad \delta_c = RX_c - TX_c - \psi$$

can solve with the simple approach to define $\delta_b/\delta_a = \delta_c/\delta_b$. The time stamps $a < b < c$ are sorted, the values can calculated on the fly:

$$\psi = \frac{(RX_b - TX_b)^2 - (RX_c - TX_c)(RX_a - TX_a)}{2(RX_b - TX_b) - (RX_c - TX_c) - (RX_a - TX_a)}$$

This approach was shown in different works at the University in Kiel [15–17]. With the estimation of the synchronization error ψ, the three travel time values $t_L \in \{\delta_a, \delta_b, \delta_c\}$ can be calculated, and for each transmission, the relative distance and the sound speed can be estimated.

In the project EDA SABUVIS 2 the Kiel approach was adopt to a simple IN-RES process (see also [18]): *OwnNode* is transmitting its timestamp $TX_{MARKown}$ inside the MARK-message shown in Fig. 3.2. *OtherNode* is transmitting its time-stamp $TX_{SNAPother}$ together with the receiving timestamp of the MARK-message $RX_{MARKother}$ inside the SNAP message. The simple two equations are

$$\{RX_{MARKother} = TX_{MARKown} + \delta + \psi, RX_{SNAPown} = TX_{SNAPother} + \delta - \psi\},$$

if the distance (in form of the delay δ) between is not changing. The offset value ψ is the time difference between the clocks from *ownNode* and *otherNode*.

OwnNode is waiting for all received SNAP messages from the *otherNode*s, the delay time to them is given by

$$\delta = \frac{1}{2}(RX_{MARKother} + RX_{SNAPown} - TX_{MARKown} - TX_{SNAPother}).$$

The synchronization offset to them is given by

$$\psi = \frac{1}{2}(RX_{MARKother} - RX_{SNAPown} - TX_{MARKown} + TX_{SNAPother}).$$

The *otherNode*s has to wait for a SNAP message from *ownNode*.

The presented techniques with passive and active range estimations helps *OwnNode* to build up own knowledge about the distances of its one-hop-neighborhood. They helps also to overcome the challenges in underwater localization, described in the article [4].

References

1. Schreiber S, Nissen I (2015) Digital distance and channel measurements with the underwater telephone ut 3000. In: Proceedings of the European conference of underwater defense technology, UDT Europe 2015
2. Cheng L, Chengdong W, Zhang Y, Hao W, Li M, Maple C (2012) A survey of localization in wireless sensor network. Int J Distrib Sensor Netw 8(12):962523
3. Nissen I (2008) Adaptive systems for mobile underwater communications with a p(oste)riori channel knowledge, first half. FWG report, 59
4. Beniwal M, Singh R (2014) Localization techniques and their challenges in underwater wireless sensor networks. Int J Comput Sci Inf Technol 5:4706–4710
5. Chen K, Zhou Y, He J (2009) A localization scheme for underwater wireless sensor networks. Int J Adv Sci Technol 4
6. Chandrasekhar V, Seah WKG, Choo YS, Ee HV (2006) Localization in underwater sensor networks: survey and challenges. In: Proceedings of the 1st ACM international workshop on Underwater networks. ACM, pp 33–40
7. Han G, Jiang J, Shu L, Yongjun X, Wang F (2012) Localization algorithms of underwater wireless sensor networks: a survey. Sensors 12(2):2026–2061
8. Ahmed M, Salleh M (2016) Localization schemes in underwater sensor network (UWSN): a survey. Indones J Electr Eng Comput Sci 1(1):119–125

9. Moradi M, Rezazadeh J, Ismail AS (2012) A reverse localization scheme for underwater acoustic sensor networks. Sensors 12(4):4352–4380
10. Cheng X, Shu H, Liang Q, Du Hung-Chang D (2008) Silent positioning in underwater acoustic sensor networks. IEEE Trans Veh Technol 57(3):1756–1766
11. Dubrovinskaya E (2012) GPS in underwater communication networks. Master thesis, Kiel
12. Dubrovinskaya E, Nissen I, Casari P (2016) On the accuracy of passive multipath-aided underwater range estimation
13. Tan H-P, Diamant R, Seah WKG, Waldmeyer M (2011) A survey of techniques and challenges in underwater localization. Ocean Eng 38(14–15):1663–1676
14. Peleato B, Stojanovic M (2007) Distance aware collision avoidance protocol for ad-hoc underwater acoustic sensor networks. IEEE Commun Lett 11(12):1025–1027
15. Konior A (2019) IN-RES protocols in underwater applications. Akademia Marynarki Wojennej, Thesis, Gdynia
16. Wachlin K (2019) Aufbau eines Feed-basierenden Anwendungssimulators für Unterwasser-MANETs im Navigations-kontext. Technische Fakultät, Kiel
17. Bülbül F, Petersen T, Recker M, Sell F, Wachlin K (2017) Fehlertoleranz bei Prozessabläufen-mit Anwendungen bei akustischen Unterwassernetzwerken. Universitätsbibliothek Kiel
18. Konior A (2020) Secure navigation of swarms and autonomous teams in the water column. Akademia Marynarki Wojennej, Thesis, Gdynia

Chapter 4
Delay/Disruption Tolerant Networking

Ronald in 't Velt, Ingrid Mulders, Arwid Komulainen, and Michael Goetz

Delay and/or Disruption Tolerant Networking (DTN) protocols are important when dealing with partitioned wireless communication clusters, i.e., when dynamic connections are often broken for longer times than the maximum allowed network latency. A combination of ad-hoc and DTN functionalities is key for underwater sensor networking where AUVs can temporarily leave the network area or where a submarine can go silent for a while [1–3].

DTN is an active research field, especially in terrestrial and space applications such as wildlife tracking, vehicular networks (mobile ad-hoc networks, MANETs), interplanetary networks, or in situations after a disaster, where the protocol enables transport of the data between groups of disconnected mobile nodes, if the communication path between senders and receivers is completely broken or disconnected for longer time periods (e.g., emission control) [4–7]. Several delay-tolerant networking protocols have been developed in the past years [1–27]. Amongst others, the (now disbanded) DTN Research Group of the Internet Research Task Force (IRTF) made some significant contributions by developing an architecture [28] and specifying protocols for DTN. The most important product of this group is the Bundle Protocol specification [29]. The DTN Working Group of the Internet Engineering Task Force (IETF) is revising and standardizing the Bundle Protocol.

R. in 't Velt · I. Mulders
Netherlands Organisation for Applied Scientific Research (TNO), The Hague, Netherlands

A. Komulainen
Swedish Defence Research Agency (FOI), Stockholm, Sweden

M. Goetz
Fraunhofer-Institut für Kommunikation, Informationsverarbeitung und Ergonomie (FKIE), Bonn, Germany

© The Author(s) 2020
D. Sotnik et al. (eds.), *Cognitive Underwater Acoustic Networking Techniques*,
SpringerBriefs in Electrical and Computer Engineering,
https://doi.org/10.1007/978-3-662-61658-1_4

4.1   Store-Carry-Forward Paradigm

Before explaining the Bundle Protocol specification, the Store-Carry-Forward paradigm [5] is introduced, a key element of DTN. In a conventional packet-switched (non-flooding) network layer such as IP, a network node that is to forward a packet consults its Forwarding Information Base (FIB) to determine the Next Hop address associated with the destination address of the packet. If a Next Hop address is found, then the packet may sit in a queue of the associated outgoing interface for a while, waiting its turn, but it will eventually be transmitted: this is referred to as Store-and-Forward. However, if no Next Hop address for the given destination is present in the FIB, then the node has no other option than to discard the packet. DTN allows a node to hold on a Protocol Data Unit (PDU) for which a Next Hop is not readily available. If the node is mobile, then its movements may bring it within communication range of the final destination of that PDU or within range of a suitable Next Hop: hence *Store-Carry-Forward*.

4.2   Bundle Protocol Specification

One of the main implementations of the Store-Carry-Forward paradigm is the Bundle Protocol specification. The idea behind the Bundle Protocol is that, in an environment with long communication delays and frequent disruptions, protocol interactions between communication endpoints should be kept to a minimum. Data to be exchanged should be grouped together with all necessary meta data and protocol information into a self-contained Protocol Data Unit, the *Bundle*. In terms of the OSI (or TCP/IP) model, the Bundle Protocol can be thought of as residing at the Application Layer, forming an overlay over the traditional protocol stack. The entity instantiating the Bundle Protocol layer at a node is called the *Bundle Protocol Agent* (BPA). The transfer of a Bundle from one BPA to another BPA is end-to-end communication from the OSI point of view, but several of such transfers in sequence may be required to convey a Bundle from its source DTN endpoint to its destination DTN endpoint, with intermediate BPAs acting as Bundle relays, see Fig. 4.1. Bundles are made up of blocks, always including a single Primary Block and a single Payload Block, and optionally one or more Extension Blocks. The Primary Block contains Bundle Protocol header information, including a Source and a Destination Endpoint Identifier (EID). These EIDs take the form of URLs, consisting of a scheme name followed by a scheme-specific part (e.g., dtn://ucomms/auv-1). Bundles can be much larger than (IP) packets, with the size of the latter typically being constrained by the Maximum Transmission Unit (MTU) of the underlying link layer technology (packet fragmentation should generally be avoided). Breaking down Bundles into segments that the Transport Layer can handle is a task for the Convergence Layer that resides below the Bundle Protocol Layer. Convergence Layers for different Transport Layer

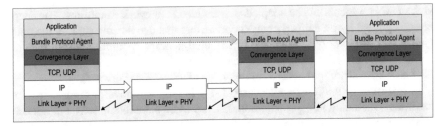

Fig. 4.1 Protocol stacks in a combined DTN Bundle-relaying and packet-forwarding configuration

protocols have been specified, e.g., for TCP [30] and for datagram-based transport [31].

In an environment where node movement patterns are largely unpredictable,[1] BPAs need to become aware of each other's presence before Bundles can be exchanged. This necessitates a mechanism for DTN *Neighbour Discovery*, which can either involve the use of a special kind of Bundles (in-band) or be based on a dedicated protocol external to the BPA (out-of-band). Moreover, when a DTN node encounters another DTN node that does not hold the Destination Endpoint for some of the Bundles that the first node is carrying, it has to decide whether or not to transfer or copy those Bundles to the other node. Essentially, it has to assess whether the encountered node is a viable candidate for getting the Bundles closer to their Destination Endpoint. This process is referred to as *DTN routing*. It should be realized, however, that this is very different from routing at the packet level. Many different DTN routing strategies can be found in literature. A distinction is made between *single-copy* and *multi-copy* schemes. The former allows only a single instance of a given Bundle to be present in the network at any time, whereas the latter uses duplication of Bundles to increase the probability of their delivery and/or decrease latency. Examples of multi-copy routing strategies are Epidemic [8, 32] and Spray-and-Wait [33], and an example of a single-copy scheme is the custody transfer mechanism [3].

4.3 DTN for Underwater Acoustic Networks

Bundle-Protocol-based DTN is developed for terrestrial and interplanetary networks, and a valid question would be to what extent it is applicable to underwater acoustic communications. The Bundle Protocol and its supporting Convergence Layer introduce additional protocol overhead, for which the time and bandwidth in the underwater acoustic network may be limited. An interesting new development to reduce communication overhead is the context-based adaptation in DTN [27], in

---

[1]In one significant area of application of DTN, space communications, disruptions and re-establishment of communication links often are predictable: it can be calculated ahead of time when a planet orbiter will emerge from the shadow of the planet, when a satellite will come into range of a ground station, etc.

which DTN protocols can be adapted to variations in network conditions. The adaptation is guided by predefined context parameters that can be collected by the node itself. DTN does not only require a (distributed) storage capability within the network, but the network should also be able to detect that a destination node cannot be reached before sending messages to that node (Store-Carry-Forward).

Furthermore, Bundle-Protocol-based DTN is targeting adverse communication conditions, including long propagation delays and frequent disruptions, but not necessarily very low transmission rates. In fact, it can be argued that the Bundle Protocol works best in environments where node encounters are of an opportunistic nature. I.e., when during contact periods a relatively high transmission rate can be used, presumably over short distances, for example using optical communication. However, the Store-Carry-Forward paradigm of DTN is deemed to be valuable in the context of underwater communications. It may be worth exploring whether this paradigm can somehow be implemented at the Network Layer instead of the Application Layer, i.e., packet-based DTN instead of Bundle-based DTN. Unfortunately, there are no known specifications to fall back on for such an alternative approach to DTN. The required technology would thus have to be developed from scratch.

Some efforts have already been made to adapt known DTN mechanisms for the underwater network environment with untrustworthy links. A first step in this direction was the ACommsNet10 trial in September 2010, performed by CMRE (La Spezia, Italy), by using new local, low-overhead, adaptive routing schemes [34]. Additionally, the capability to exchange data between separated network clusters with AUVs was tested within a German national WTD 71/FKIE cooperation. For this purpose, the network protocol was extended, among other enhancements, by a so-called postman functionality. The protocol extension was tested with two SeaCat AUVs and three bottom nodes during a sea trial near Bornholm in November 2014 and in Summer 2017 in the North Sea. GUWAL was used as application language, which was already used in the RACUN demonstration and was only extended by DTN network control packets, e.g., for the postman handshake. An advantage of GUWAL are the timestamps within the packets, which can be used to decide if a delayed packet is still valid and should be exchanged with a postman or should be dropped.

4.4   Data Muling

A commonly used application for DTN is Data Muling. Data muling is the activity of transporting data between (static) nodes using one or more mobile nodes. In Underwater Sensor Networks (UWSNs), data muling usually refers to an AUV transporting data between nodes, or clusters of nodes, between which there are no (long-range) communication links. Typical scenarios that employ data muling are postman scenarios, where communication between disjoint parts of a network is assisted by a mobile postman, data offloading from static sensor nodes to mobile nodes, and data offloading from mobile (inspection) nodes to static gateway nodes (wireless docking). Data offloading is typically performed using some form of high-speed communication technology. Optical communication is commonly used [35–37], and another option is to use a high-frequency acoustic link.

Some prerequisites for efficient data muling are: AUV localization of and distance estimation to sensor nodes using, for instance, the methods presented in the previous chapter; high-speed communication links, high-frequency acoustic or electromagnetic; and protocol stacks designed to handle intermittent links (delay/disruption-tolerant-networking). Depending on the purpose of the data mule and the scenario in which it is used, efficient route planning in order to visit sensor nodes in an energy-efficient manner can also be an important requisite. In [38], Hollinger et al. treated route planning as a traveling salesman problem, with the problem formulation modified to account for the unreliable communication links.

In [36], the AUV is equipped with a down-facing camera and is assumed to have low-accuracy maps of sensor node locations. The AUV also surfaces occasionally to correct drift in its position estimate. The AUV is tasked with visiting all stationary nodes to offload data. The sensor nodes are equipped with low-rate acoustical modems which can be used for signaling events and beaconing, but this is not demonstrated in the paper. When the AUV is in the vicinity of a node, it processes the images from the camera to locate and hover over the sensor node. When the AUV has started hovering it uses optical communication to offload data before moving to the next node. The system is demonstrated in a pool containing three sensor nodes and one AUV.

The system demonstrated in [37] uses acoustic beaconing to let the AUV locate the sensor nodes using either a stochastic gradient descent approach or a particle filter. Both methods are shown to successfully localize the sensor node, with the particle filter performing best. The AUV is said to need only a rough estimate of the sensor node position, with an error margin similar to the range of the acoustic beacon. The sensor node sends acoustic beacons every six seconds and constantly streams optical data. When a packet is successfully received by the AUV over the optical link it switches from using the acoustic beacons to the optical signal to stay hovering over the sensor node.

In [39], the medium access problem that occurs when a data mule visits a cluster of nodes is examined. More specifically, the authors compare random access schemes with a proposed polling-based scheme called *UW-polling*. The protocol consists of three parts: a neighborhood discovery phase where the AUV determines which nodes to poll, a data prioritization phase where the nodes communicate what data to send and finally the AUV polls the nodes in an order determined in the previous phase. UW-polling is compared to random-access schemes, CS-Aloha and Distance-aware collision avoidance protocol (DACAP), in terms of throughput, packet delivery ratio, delay and energy efficiency. The CS-Aloha protocol is modified in the sense that the AUV transmits trigger packets which triggers the Aloha procedure in the sensor nodes for a given time. The modification prevents sensor nodes to transmit blindly when the AUV is out of range. At low source power levels UW-polling offers benefit in terms of robustness compared to the other schemes, however, for higher source levels, the performance is similar to CS-aloha. In [40], UW-polling is compared to another protocol proposed by the same authors called *U-Fetch*, in which sensor nodes forward data to cluster heads, which are appointed for communicating with the AUV. It is shown that U-Fetch provides lower latency than UW-polling, at the cost of lower packet delivery ratio.

A multi-modal data mule scenario using a combination of acoustic and optical communications is analyzed in [41]. The system uses the previously mentioned CS-Aloha-Trig protocol with a fixed physical-layer modality during each trigger period. Between trigger periods, the AUV can decide to switch between acoustic and optical communication, based on the received power for each modem. Network simulations using the DESERT framework [42], extended to model multi-modal communications, are used to determine the throughput in different water conditions. To handle difficult, turbid conditions, the data mule needs to stand still longer at each node in order to correctly determine when to switch physical-layer modality.

References

1. Anitha S, Micheal G (2013) An evaluation of different routing schemes in dtn. Int J Comput Sci Mob Comput 2(5):41–49
2. Desai CB, Pandya VN, Hadia SK (2013) A survey on knowledge based classification of different routing protocols in delay tolerant networks. Int J Comput Sci Mob Comput 2(3):83–88
3. Yu Y, Chen X (2013) Research on custody transfer service in delay tolerant network. J Netw 8(8):1713–1719
4. Omidvar A, Mohammadi K (2014) Particle swarm optimization in intelligent routing of delay-tolerant network routing. EURASIP J Wirel Commun Netw 147
5. ZhuK (2014) Social-based Data Routing Strategies in Delay Tolerant Networks. PhD thesis, University of Göttingen
6. Ono M, Sawai K, Suzuki T (2014) Development of rest facility information exchange system by utilizing delay tolerant network. Int J Adv Comput Sci Appl 5(2):83–89
7. Wang S-Y, Torgerson JL, Schoolcraft J, Brenman Y (2014) The deep impact network experiment operations center monitor and control system. http://hdl.handle.net/2014/44657
8. Marandi A, Faghi Imani M, Salamatian K (2014) Practical bloom filter based epidemic forwarding and congestion control in dtns: a comparative analysis. Comput Commun 48:98–110. Elsevier
9. Ramesh S, Ganesh Kumar P (2014) Bcr routing for intermittently connected mobile ad hoc networks. Int J Eng Technol 6(1):66–74
10. Zhang Z, Jin Z, Ma M (2014) Ccs-dtn: clustering and network coding-based efficient routing in social dtns. Sensors 15(1):285–303
11. Papaj J, Dobos L (2014) Trust based algorithm for candidate node selection in hybrid manet-dtn. Adv Electr Electron Eng 12(4):271–278
12. Jain S, Kishore N, Chawla M (2014) Soares VNGJ (2014) Composite mechanisms for improving bubble rap in delay tolerant networks. J Eng 1:1–7
13. Garcia G, Robles S, Sanchez A, Borrego C (2014) Information system for supporting location-based routing protocols. RECSI, pp 203–208
14. Moreira WA Jr (2014) Social-aware opportunistic routing. PhD thesis, Universidade de Aveiro
15. Saha S, Verma R, Saika S, Paul PS, Nandi S (2014) e-one: enhanced one for simulating challenged network scenarios. J Netw 9(12):3290–3304
16. Zou S, Wang W, Wang W (2013) A routing algorithm on delay-tolerant of wireless sensor network based on the node selfishness. EURASIP J Wirel Commun Netw p 212

17. Xie LF, Chong PHJ, Guan YL (2013) Routing strategy in disconnected mobile ad hoc networks with group mobility. EURASIP J Wirel Commun Netw p 105
18. Fiems D, Altman E (2009) Markov-modulated stochastic recursive equations with applications to delay-tolerant networks. Technical Report RR-6872, INRIA. An optional note
19. Liang H (2013) Resource management in delay tolerant networks and smart grid. PhD thesis, University of Waterloo
20. Poonguzharselvi B, Vetriselvi V (2012) Trust framework for data forwarding in opportunistic networks using mobile traces. Int J Wirel Mob Netw 4(6):115–126
21. Ramesh S, Ganesh Kumar P (2013) Spray and wait routing with agents in intermittently connected manets. Int J Eng Technol 6(1):66–74
22. Wang S, Ma R (2013) Name: A naming mechanism for delay/disruption-tolerant network. Int J Comput Netw Commun 5(6):231–241
23. Fowjiya S, Udhayachandrika A, Kathirvel A (2013) Architectural overview of delay tolerant network. Int J Eng Sci Res Technol 2(10)
24. Gao L, Li M, Zhou W, Shi W (2013) Anonymous data forwarding in human associated delay tolerant networks. In: IEEE 33rd international conference on distributed computing systems workshops, July 2013. Philadelphia (PA), USA
25. Gao L, Li M, Bonti A, Zhou W, Yu S (2013) Multi-dimensional routing protocol in human associated delay-tolerant networks. IEEE Trans Mob Comput 12(11):2132–2144
26. Bindra HS, Sangal AL (2012) Need of removing delivered message replica from delay tolerant network - a problem definition. Int J Comput Netw Inf Sec 4(12):59–64
27. Petz A (2012) Context-based adaptation in delay-tolerant networks. PhD thesis, University of Texas, Austin, USA
28. Cerf V, Burleigh S, Hooke A, Torgerson L, Durst R, Scott K, Fall K, Weiss H (2007) Delay-tolerant networking architecture. RFC-4838, IRTF network working group
29. Scott K, Burleigh S (2007) Bundle protocol specification. RFC-5050, IRTF network working group
30. Demmer M, Ott J, Perreault S (2014) Delay-tolerant networking tcp convergence-layer protocol. RFC-7242, IRTF network working group
31. Kruse H, Jero S, Ostermann S (2014) Datagram convergence layers for the delay- and disruption-tolerant networking (dtn) bundle protocol and licklider transmission protocol (ltp). Technical Report RFC-7122, IRTF network working group
32. Lu X, Hui P (2010) An energy-efficient n-epidemic routing protocol for delay tolerant networks. In: Fifth IEEE international conference on networking, architecture and storage, Macau SAR, China
33. Spyropoulos T, Psounis K, Raghavendra CS (2005) Spray and wait: an efficient routing scheme for intermittently connected mobile networks. In: Proceedings of 2005 ACM SIGCOMM workshop on delay-tolerant networking, Philadelphia (PA), USA
34. Merani D, Berni A, Potter J, Martins R (2011) An underwater convergence layer for disruption tolerant networking. In: Baltic congress on future internet and communications, Riga, Latvia
35. Vasilescu I, Kotay K, Rus D, Dunbabin M, Corke P (2005) Data collection, storage, and retrieval with an underwater sensor network. In: Proceedings of the 3rd international conference on embedded networked sensor systems, SenSys '05. ACM, New York, NY, USA, pp 154–165
36. Dunbabin M, Corke P, Vasilescu I, Rus. D (2006) Data muling over underwater wireless sensor networks using an autonomous underwater vehicle. In: Proceedings 2006 IEEE international conference on robotics and automation, 2006. ICRA 2006, pp 2091–2098
37. Doniec M, Topor I, Chitre M, Rus D (2013) Autonomous, localization-free underwater data muling using acoustic and optical communication. Springer International Publishing, Heidelberg, pp 841–857
38. Hollinger GA, Choudhary S, Qarabaqi P, Murphy C, Mitra U, Sukhatme GS, Stojanovic M, Singh H, Hover F (2012) Underwater data collection using robotic sensor networks. IEEE J Sel Areas Commun 30(5):899–911
39. Favaro F, Casari P, Guerra F, Zorzi M (2012) Data upload from a static underwater network to an auv: polling or random access? In: 2012 Oceans - Yeosu, pp 1–6

40. Favaro F, Brolo L, Toso G, Casari V, Zorzi M (2013) A study on remote data retrieval strategies in underwater acoustic networks. In: 2013 OCEANS - San Diego, pp 1–8
41. Campagnaro F, Favaro F, Guerra F, Calzado VS, Zorzi M, Casari P (2015) Simulation of multimodal optical and acoustic communications in underwater networks. In: OCEANS 2015 - Genova, pp 1–7
42. Casari P, Tapparello C, Guerra F, Favaro F, Calabrese I, Toso G, Azad S, Masiero R, Zorzi M (2014) Open source suites for underwater networking: Woss and desert underwater. IEEE Netw 28(5):38–46

Chapter 5
Multi Topology Routing

Michael Goetz

One of the main challenges in an adaptive network is using multiple disjunct frequency bands (multi band approach). Additionally to currently used one band (e.g. a lower one), a higher frequency band can be implemented. This will have several benefits like higher data rates, lower ambient noise, and smaller transducers. Though, for the network protocol arise new challenges due to more complex routing. The high-frequency band has a significantly lower range, which leads to less direct links and therewith another topology compared to the low-frequency band. This phenomenon is named "Multi-Topology". Additionally, radio links between surface nodes like buoys or ships may be present in the network. These radio links can also use different frequency bands like HF, LTE or WLAN resulting in different ranges and data rates. Figure 5.1 shows different topology views for low and high-frequency underwater acoustic links and the radio link in air above the surface. Obviously, none of the topologies connect all of the nodes, i.e., high frequency can communicate within a barrier only, or low frequency cannot communicate with AUVs. For this reason, the technologies must be combined and a route can cover different transmission methods as shown in Fig. 5.2. The figure shows also multiple possible paths for routing. One path uses the buoys to transmit data via air, which may be more reliable and faster than the alternative route over low-frequency underwater communication via the second barrier (shown as dotted line). Please note, yet it is unknown if all nodes can at least receive both low and high-frequency transmissions, such that at least a one-directional link is available. In this way, a low and a high-frequency transmitter may communicate with each other, each in its own frequency band.

For now, there are only a few publications about routing over high and low-frequency bands in underwater networks. Survey papers like [1] address various challenges for routing protocols, but no one mentions the use of multiple frequency bands with different ranges in underwater networks. But, there is a proposal for the

M. Goetz
Fraunhofer-Institut für Kommunikation, Informationsverarbeitung und Ergonomie (FKIE),
Bonn, Germany

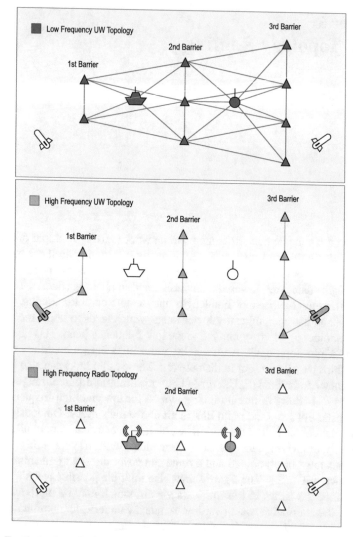

Fig. 5.1 Topology views for low and high-frequency underwater acoustic links and radio link in air above the surface

combination of wireless radio and underwater acoustic communication in [2]. Liu et al.introduce a novel on-surface wireless-assisted opportunistic routing (SurOpp) for underwater sensor networks. Within this work, multiple buoys were deployed to act as sinks for underwater nodes. The bottom nodes use acoustic communication to deliver multiple sensor packets to the surface. At the surface, radio links are used to forward the packets with an opportunistic routing strategy. This means, all nodes overhear the communication of others and will reject its own forwarding if another node already forwarded the packet. Unfortunately, they assume that all surface nodes

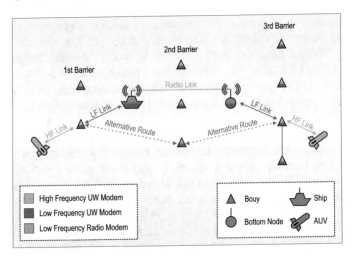

Fig. 5.2 Multi Topology Routing over different communication technologies and alternative routes represented as dotted lines

are connected with each other, and therefore the underwater nodes do not have to worry about which surface nodes receive the data. As a result, there is no routing in the underwater domain necessary. Nevertheless, in the RACUN project, it could also be shown that the total packet delivery ratio could be enhanced by 10– 20% if surface nodes are connected with each other [3].

In field experiments of TNO and FFI [4], the first step for dual-frequency-band communication was done. They used a low (4–8 kHz) and a high-frequency band (26–30 kHz) in a setup of three NILUS bottom nodes, an AUV and a base station on a boat. The bottom nodes and the base station used hydrophones, allowing them to decode both low and high-frequency transmissions. But, the AUV could only communicate within the high-frequency band. Additionally, the bottom nodes were equipped with low-frequency transducers only, that means they could receive but not transmit messages in the high-frequency band. Therefore, communication from the bottom nodes to the AUV was only possible through the base station. This was a first step for dual-band communication by combining two different frequency bands with different transducers. Even though, the focus was not set on the routing yet. Each node was preconfigured and knew if the destination should be addressed in the low or high-frequency band. The routing issue gets much more complicated in multi-hop scenarios with stationary and mobile nodes, which are roaming around like in ISO IEC 30140.

A paper of the University of Rome addresses the medium access control (MAC) for multi-band underwater networks [5]. The aim was to dynamically adapt the frequency band based on the current noise power spectral density and the signal-to-noise ratio in the water column. The protocol is named Noise-Aware MAC protocol (NAMAC) and introduces a cooperative strategy to decide which frequency band should be used for communication. They also take into account borders where networks operate on different frequency bands. The heterogenity of the network with various nodes supporting different frequency bands will also be one of the main challenges in the project EDA SALSA. However, NAMAC does not care about routing and neither about completely different frequency bands. They assumed a single transducer within a frequency band of 5–10 kHz which is divided into multiple subbands.

The optimal route selection is also a key challenge in wireless networks with different radio technologies within a network. Most of the routing protocols use a specific cost metric like hop count, bandwidth, latency or a combination of these. In [6], the authors introduce a new design method to select optimal routes in wireless networks and consider also Ethernet links within the routes. The idea is to use multiple routing tables for different traffic classes. VoIP requires low latency link, whereas an FTP application wants a high throughput. This can lead to different routes, which will be covered by different routing tables. In SALSA, it is conceivable that we have applications with high data rates and large packet sizes and applications with a high demand on robust links and reliable communication. Therefore, optimal route selection should be also considered for the routing.

References

1. Sarao P, Chattu K, Swapna Ch (2018) Routing issues and challenges in underwater wireless sensor networks. Int J Comput Sci Eng 6:02
2. Liu M, Ji F, Guan Q, Yu H, Chen F (2016) On-surface wireless-assisted opportunistic routing for underwater sensor networks
3. Nissen I, Goetz M, Schreiber S (2015) Secure underwater coordination of manned and unmanned platforms. Naval Forces. NF IV, pp 57–58
4. Dol H, Colin M, van Walree P, Otnes R (2018) Field experiments with a dual-frequency-band underwater acoustic network. In: 2018 fourth underwater communications and networking conference (UComms), pp 1–5. IEEE
5. Pescosolido L, Petrioli C, Picari L (2013) A multi-band noise-aware mac protocol for underwater acoustic sensor networks. In: 2013 IEEE 9th international conference on wireless and mobile computing, networking and communications (WiMob), pp 513–520
6. Wang M, Davidson SA, Chuang YS (2013) A design method to select optimal routes and balance load in wireless communication networks. In: MILCOM 2013 - 2013 IEEE military communications conference, pp 916–921

Chapter 6
Autonomous Ad Hoc Networks

Roald Otnes and Ivor Nissen

To realize centralized communication strategies underwater is very cost-intensive and not feasible due to the partly scattered network nodes. Each node must be able to distribute its messages directly. The communication should work without configuring all parties to a certain communication path. Each node should adapt its operational addresses and network parameters to the existing network cluster for itself. To do this, the participants must exchange some initial information. In radio transmissions, there are usually uniform standards for the communication procedure. Thus everyone speaks the same language. Underwater, however, manufacturer-specific procedures and codings are usually used. In order for the participants to understand each other, they need a procedure with which they can arrange particular link parameters.

6.1 JANUS

The JANUS standard [1, 2] is a simple multiple-access acoustic protocol designed and tested by the NATO Centre for Maritime Research and Experimentation (CMRE) with the collaboration of academia, industry and government. It was developed with the initial contact application in mind based on the impressions of the *KURSK* submarine disaster in 2000. The original idea is to allocate time for other transmissions (or for silence for other sonar applications). The first contact aspect of JANUS is modeled after VHF channel 16[1] in terrestrial applications. After initial contact is

---

R. Otnes
Norwegian Defence Research Establishment (FFI), Kjeller, Norway

I. Nissen
Wehrtechnische Dienststelle für Schiffe und Marinewaffen, Maritime Technologie und Forschung (WTD 71), Eckernförde, Germany

---

[1]Channel 16 (156.8 MHz) is a marine VHF radio frequency designated as an international distress frequency.

© The Author(s) 2020
D. Sotnik et al. (eds.), *Cognitive Underwater Acoustic Networking Techniques*,
SpringerBriefs in Electrical and Computer Engineering,
https://doi.org/10.1007/978-3-662-61658-1_6

established, one can either continue communicating through JANUS, or switch to a more capable waveform common to the communicating parties.

The JANUS standard itself only defines the PHY and MAC layers but not the applications, and hence does not say anything about initial contact/node discovery. However, CMRE has defined applications for JANUS in detail, including first contact, in a journal paper [3].

Sec. III-A of [3] describes CMRE's proposal for the first contact application with JANUS, which Petroccia, Alves and Zappa have tested experimentally in 2015 in a setup where the nodes had common capabilities of transmitting JANUS and Evologics[2] modulation, and switched to Evologics after establishing first contact through JANUS. The initial contact protocol has three phases, here paraphrased from [3]:

1. *Node discovery:* A node initiates the initial contact protocol by periodically broadcasting a discovery request. Another node that receives the discovery request can send back a reply (after a random delay) to make the requesting node, and other nodes in the vicinity, aware of its presence. After this procedure, the nodes are aware of each other's presence.

2. *Communication capability discovery:* In this phase, a pair of nodes exchanges messages to agree on a common modulation scheme. One node initiates the process by sending a communications request message to the other node. This message includes a list identifying which modulation schemes it supports. The other node replies with a response listing which modulation schemes the two nodes have in common. After this process, both nodes are aware of which modulation schemes they have in common.

   If either of the packets is lost and the requesting node does not receive any response, it will retransmit the request when a timeout expires.

3. *Language switching:* Here, the requesting node from the previous phase decides which common modulation scheme ("language") to use with the other node, and sends it a language request message. The other node sends a response where it accepts or rejects the request. If it accepts, it also sends back values for timeouts which determine when to switch back to JANUS.

   If either of the packets is lost and the requesting node does not receive any response, it will switch back to JANUS after a timeout expires. The other node will also switch back to JANUS if it does not receive any data in the "new language" before a timeout expires.

The three phases are illustrated in Fig. 6.1, for the case of no lost messages. Note that [3] does not propose any scheme by which more than two nodes can switch language through one message exchange, hence each pair of nodes in the network would have to exchange the messages of phase 2 and 3. This would have to be reworked, e.g., for the case of one new node entering a network where many existing nodes have already agreed upon a common non-JANUS language.

---

[2]EvoLogics GmbH is a high-tech enterprise which design and manufacture wireless underwater communication systems.

Fig. 6.1 JANUS initial
contact protocol exchanges
(illustrations from [3]).
Top: Node discovery phase.
Center: Communication
capability discovery phase.
Bottom: Language switching
phase. D-Req/D-Resp =
Discovery request and
response. C-Req/C-Resp =
Comms request and
response. L-Req/L-Resp =
Language switch request and
response. MCS =
Modulation and coding
scheme

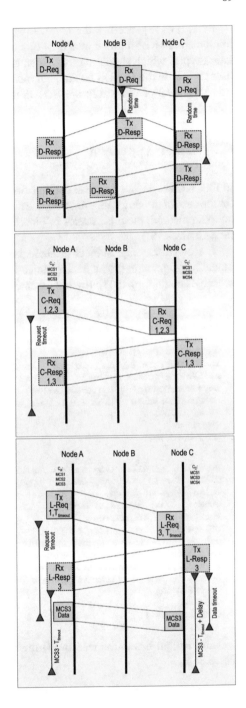

Table II of [3] specifies the JANUS packets that were used for the experimental implementation of initial contact. 24 bits were allocated to specify different modulation types, which should be more than enough. They used a currently open class user ID (13) in the JANUS format, but mention the possibility to "populate the class user ID allocation table in future JANUS standard revisions".

6.2 Kieler Approach

In this subsection, we propose another first-contact approach building on JANUS, for the case of moving nodes entering the network. Based on the specified frequency-hopping method, the mini stories in Figs. 6.2 and 6.3 are described in ([4], Chap. 6, First Contact, [5]).

The corresponding Subject-oriented business process management (S-BPM)-modeling graph with the intercommunication is given in Fig. 6.4, with three defined messages containing the following parameters:

```
MESSAGE : "Hello, I am the new one"
Class User ID := 4, Application type := 0
buildInPayload JANUS   :=  [ (34bit)
ownName (last nibble of name)                             4 bit
knownNeighbors (my neighborhood) array[16] of boolean     16 bit
mainPhyConfiguration (modulation and coding)              4 bit
secondPhyConfiguration (modulation  and coding)           4 bit
NetConfiguration  (link and networklayer)                 3 bit
ChannelModel                                              3 bit ]
Checksum                                                  8 bit
CargoHold :=[]
```

```
MESSAGE :  "Hello, I am not compatible!"
Class User ID := 4, Application type := 1
buildInPayload JANUS   :=  [ (34bit)
ownName (last nibble of name)                             4 bit
yourCRC (CRC of message 0)                                8 bit
otherNeighborName1  (current)                             4 bit
otherNeighborName2  (second current)                      4 bit
mainPhyConfiguration (modulation and coding)              4 bit
secondPhyConfiguration (modulation  and coding)           4 bit
NetConfiguration  (link and networklayer)                 3 bit
ChannelModel                                              3 bit ]
Checksum                                                  8 bit
CargoHold :=[]
```

Fathi Bülbül has introduced in [4] the new concept of a translator subject (see Fig. 6.3c):

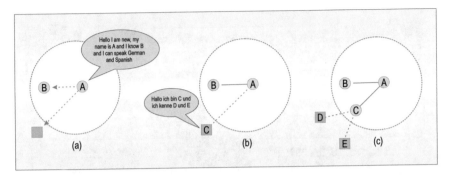

Fig. 6.2 A new subject A connects a MANET cluster and is transmitting the first contact message (**a**). The subject C is compatible and is answering in a high level language (**b**). The mini story was finished positive (**c**) [4]

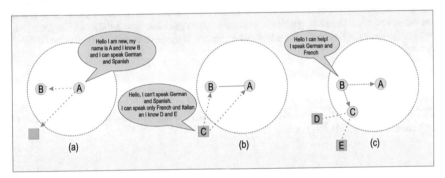

Fig. 6.3 A new subject A connects a MANET cluster and is transmitting the first contact message (**a**). The subject C is NOT compatible and is answering in a JANUS language (**b**). The subject B can help as translator between A and C, the mini-story finished positive (**c**) [4]

```
MESSAGE :   "Hello,  I can help (Translator)"
Class User ID := 4, Application type := 2
buildInPayload JANUS   :=  [ (34bit)
ownName (last nibble of name)                           4 bit
yourCRC (CRC of message 1)                              8 bit
LanguageMessage 0,  PhyConfiguration Node A             4 bit
LanguageMessage 1, PhyConfiguration Node B              4 bit
NetConfiguration  (link and networklayer)              3 bit
ChannelModel                                            3 bit ]
Checksum                                                8 bit
CargoHold :=[]
```

All three first-contact messages are included in the Communication View Diagram of S-BPM (Subject-oriented business process management) (Fig. 6.4). If *ownNode* and *otherNodes* use the same configuration, a translation is not needed; one example for this case of a higher language under the water surface after the first contact process, is GUWAL (Generic Underwater Application Language). It uses a standard

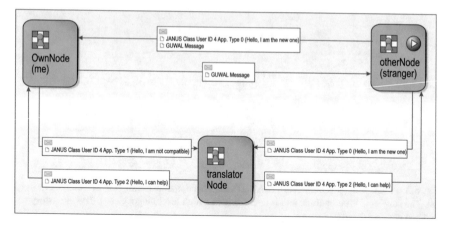

Fig. 6.4 *Source* [4], Communication view of the first contact with three subjects. If *ownNode* and *otherNodes* use the same physical and network configuration (speak the same language), the first-contact-process does not need a translator subject. They can switch to the common GUWAL message

length of 128 bits, which can be extended up to 2 KiB. GUWAL supports various applications, like GPS positioning, command & control of underwater devices or text messaging (SMS/chat).

6.3 Address Assignment

Address assignment refers to the problem of assigning addresses to nodes deployed in a specific deployed network, when the number of bits in the unique MAC address is too large to be used in every header field of the network protocol. MAC addresses should in principle be unique "world-wide", and therefore require many bits. On the other hand, a header in the network protocol may include several address fields (e.g., source, destination, last hop), and the number of bits in the address should therefore be kept as low as possible.

The assigned addresses should ideally be unique across the deployed network, but the GUWMANET protocol only requires them to be unique in a two-hop neighbourhood of each node (as discussed at the end of this section).

In IP networks (cabled or radio), the address assignment problem is commonly handled by DHCP (Dynamic Host Configuration Protocol). In that case, the Ethernet MAC addresses are 48 bits, and the assigned IPv4 addresses are 32 bits. The DHCP protocol relies on a central DHCP server in each subnetwork, and is designed for the IP protocol. It is therefore not deemed suitable for our case of ad hoc underwater networks.

The literature on address assignment for underwater networks is sparse, but we have found a few relevant references. In [6], Agrawal *et al.* propose an address assignment and address resolution protocol. In this protocol, a node joining a network computes a hash of its unique "node name" (which plays the role of MAC address) and uses the hash as its initial network address which it transmits as an "initial hash packet" to check for address conflicts. A neighbouring node, which receives this, checks if the proposed address of the new node is already in use in the network according to their locally stored address table. If no, it accepts the address and transmits an updated version of its address table which the joining node and other neighbours can store. If yes, it proposes a new address to the joining node which is the next available address number above the proposed address. As long as the involved already-deployed nodes have correct address tables for the network when the procedure starts, the end state should be that the joining node has received an address not already in use in the network. [6] does, however, not describe how the protocol handles lost packets and retransmissions or duplicate address requests.

In [7], Petroccia proposes an address assignment and topology discovery protocol called DIVE (Distributed Id assignment and topology discoVEry). For address assignment, each node generates a random key value, and when the DIVE protocol is used in the topology discovery phase the values of random keys are communicated throughout the network, so after topology discovery all nodes should have the same list of random key values. There is also a procedure to detect and recover from duplicate random key values. Each node will then sort the list of key values incrementally, so the lowest key value is address 1, the next lowest key value is address 2, and so on. As long as all nodes have the same list of random keys, they will also generate the same list of network addresses. When a new node joins the network, it will generate a random key value and reinitiate the DIVE protocol to broadcast its key value in the network. The other nodes do not change their key values, but after the DIVE protocol has finished, again they will reassign the network addresses based on the new list of random key values existing in the network. One apparent vulnerability which is not addressed in [7], is that if the list of random keys is not exactly identical in all nodes (e.g. due to packet losses), there will be chaos as different address lists result locally in different nodes as they incrementally sort different lists of keys.

A possible new idea based on the references mentioned above is: Each node can have a unique MAC address and additionally pick a random "ID" (similar to the random key above), which is a small number of bits. Then it computes a short hash of the concatenation of MAC and ID. The address used in the network will be the concatenation of the ID and the short hash. This should reduce the probability of address conflicts (as different nodes most likely will produce different hashes for the same ID). Some procedure would still have to be designed to detect and recover from address conflicts.

The approach in GUWMANET is somewhat different from what is discussed above, and is described by M. Goetz and I. Nissen in [8]. There, the network addresses assigned at the application layer are not necessarily unique, as they can, e.g., be used for multicast to all nodes with same address. Instead, the routing protocol locally uses "nicknames", which are 5-bit addresses which are unique in the two-hop neighbourhood of each node. The mechanism to assign nicknames is described in [8] as.

"The idea is to overhear at first the network traffic for a listen period T_L. If during this period communication takes place in the neighborhood, the node becomes acquainted with its neighbors and learns their nickname passively. Additionally, it gets information about its 2-hop neighborhood, because the network header of GUW-MANET includes beside the transmitter's nickname also the nickname of the last hop, as described later. After T_L the node knows a subset of its 2-hop neighborhood and chooses randomly a presumably free local nickname. Now it transmits to its local neighborhood a nickname notification (NN) parcel to indicate its presence and nickname choice. The chosen nickname is included in the network header as source address."

References

1. NATO. Digital underwater signalling standard for network node discovery & interoperability, March 2017. Standardization Agreement STANAG 4748 Edition 1
2. NATO. Digital underwater signalling standard for network node discovery & interoperability, March 2017. NATO Standard ANEP-87 Edition A Version 1
3. Petroccia R, Alves J, Zappa G (2017) JANUS-based services for operationally relevant underwater applications. IEEE J Ocean Eng 42:994–1006
4. Bülbül F, Petersen T, Recker M, Sell F, Wachlin K (2017) Fehlertoleranz bei Prozessabläufen-mit Anwendungen bei akustischen Unterwassernetzwerken. Universitätsbibliothek Kiel
5. Nissen I, Kramer F, Thalheim B (2019) S-BPM. Information Modelling and Knowledge Bases vol 312, p 137
6. Agrawal R, Chitre M, Mahmood A (2016) Design of an address assignment and resolution protocol for underwater networks. In: Oceans Asia, IEEE
7. Petroccia R (2016) A distributed ID assignment and topology discovery protocol for underwater acoustic networks. In: Third underwater communications and networking conference (UComms). IEEE
8. Goetz M, Nissen I (2012) GUWMANET-multicast routing in underwater acoustic networks. In: 2012 military communications and information systems conference (MCC), pp 27–42. IEEE

Chapter 7
Summary

Michael Goetz and Dimitri Sotnik

The declared objective is to bring the Underwater networks to a higher level of flexibility. The resulting network should be able to extend itself autonomously with (mobile) nodes of co-operating navies, implying the need for a smart-adaptive multi-band multi-lingual delay-tolerant network. Smart adaptivity concerns, for example configuration and modulation of frequency bands, data rate, communication range, etc.

- A common language spoken by all underwater acoustic modems in a common frequency band.
- A mechanism for negotiation (criteria: type of communication needed, channel/environmental conditions) of another more efficient common protocol after first contact.
- Availability of common (library of) 'state-of-the-art' protocols suited to heavy communication tasks in relevant military scenarios.
- Dynamic address allocation and identification mechanism in underwater acoustic networks.
- Delay/disruption-tolerant networking in relevant military scenarios.

In this deliverable, a State-of-the-Art study has been done, to achieve a good and feasible network adaptivity. The current research documents have been read and the most relevant technologies presented. It is not only necessary to work on underwater studies, but also on current terrestrial network proposals like the 5G trend. The topics are separated in Multi-topology routing, Delay/Disruption tolerant networking, Initial contact, Address assignment, Distance estimation, Data Muling and Adaptive Network. The aim is to find the sub-processes that can be realized in underwater networks, as each of them has its own characteristics and progress in research.

M. Goetz · D. Sotnik
Fraunhofer-Institut für Kommunikation, Informationsverarbeitung und Ergonomie (FKIE),
Bonn, Germany

© The Author(s) 2020
D. Sotnik et al. (eds.), *Cognitive Underwater Acoustic Networking Techniques*,
SpringerBriefs in Electrical and Computer Engineering,
https://doi.org/10.1007/978-3-662-61658-1_7

This state-of-the-art study summarizes the current works about cognitive network-layer methods in underwater networks. The aim of this study was getting familiar with relevant research already done in this area, and finding technology gaps needed to be solved for underwater IoT. The report shows that the issue of adaptivity has recently become increasingly important. Much can be abstracted from non-underwater studies, but needs to be adapted due to the severe limitations of underwater networks. In the beginning, a common definition of important designations is given to create a common understanding of cognitive radios and the differences between inter- and intra-node knowledge. The literature study is divided into the different relevant topics.

At first, Multi-Topology-Routing (MTR) showed to be a mainly unexplored topic in underwater environments. At the best of our knowledge, there is not much research about parallel routing in a low- and high-frequency band, consequently resulting in different connectivity graphs. The idea of exploiting the wireless channel has been also discussed to improve the overall Packet-Delivery-Ratio (PDR).

Following, the current research results in Delay/Disruption Tolerant Networks (DTN) has been presented. DTN technologies are promising in current terrestrial and space applications. The specifications of the DTN Research Group of the Internet Research Taskforce (IRTF) have been introduced, like the store-and-forward paradigm and the Bundle Protocol Agent (BPA). Nevertheless, the literature study showed that these technologies can only partially transferred to the underwater domain, due to the limited resources and low bandwidths.

For the first contact process, first works were already done for the underwater domain, like JANUS in the NATO context. Three main phases have been identified: Node discovery, communication capability discovery and language switching. Besides this, an S-BPM model of another approach was presented, including the introduction of translator nodes if two nodes do not share any common language. Here, an example for JANUS packets with 34-bit internal payload are defined, which can be used as a startup for. The network should have a full ad-hoc capability. This requires an automatic address assignment for internal network addresses, which may be a part of the first contact process. In the address assignment chapter, different methods are discussed to allocate unique addresses within the network. The GUWMANET approach is also presented, which negotiates unique addresses in the two-hop neighborhood.

Distance estimation and data muling are further topics for cognitive radios under water which have to be solved at the network layer. The distance estimation is important to decide when to switch the frequency and for the postman or data muling functionality. Distance estimation is very difficult in the underwater domains, as GPS does not exist and nodes typically are not synchronized in time. A distinction between target/source localization and node self-localization was done. For SALSA, a handshaking process like the IN-RES protocol may be suitable, but also a passively self-localization approach exploiting the multi-path propagation in underwater channels.

For data muling, some studies with AUVs are presented that combine acoustical and optical communication. Another idea is to use a Medium Access Control

(MAC) method with collision avoidance to transmit large data sets at once in a shared medium.

Additional to the immediate adaptivity at the modem, there exist cognitive and adaptive decisions that may be made by the network layer. It has been discussed that profiles should be declared to configure the modems. These profiles need a solid basis to be drafted. In the current State-of-Art, mostly SNR is used with Feedback Channels, as well as Channel Impulse Response. Most of them have been only tested in simulation or under laboratory conditions. Since one of the highest priorities in underwater communication is efficiency, as little unnecessary messages or signals as possible should be sent. Therefore, the approach of self-awareness (Sect. 1.4), which can use echos for Intra-Knowledge could be used profitably. It has to be included at least in a global or few separate *Performance metrics*, which have to be found in this project. But the approaches of Sect. 2.5 to use a *particle filter* or a *Markovian model* can be adopted.

Based on the present literature study for adaptive modulations, it is learned that most efforts so far have focused on adaptation of the source level and the data rate. A number of other possible adaptions, such as swapping between coherent and incoherent modulation schemes and redistribution of power across carriers, is discussed in Sect. 2.1.

As is explained in Sect. 2.2, the modem settings can be optimized based on user/scenario requirements and application settings, e.g., maximum source level, (unfragmented) message size, max. relative speed and the required Quality-of-Service. Settings can also be optimized based on the environmental conditions, where we can distinguish between physical descriptors, such as input SNR, ambient noise level, delay/Doppler spread, channel coherence time and inter-node ranging measurements, and system descriptors, such as output SNR, Doppler shift, BER estimation and degree of clipping.

In most found publications, the synchronization of adaptions between transmitter and receiver is realized by sending separate feedback messages, which is flexible but vulnerable and increases the network latency. A more limited but robust alternative is to use preamble encoding, at the price of additional overhead per message. Furthermore, also the option of classification of received messages is discussed in Sect. 2.3

Finally, it is investigated in Sect. 2.4 what has been reported on how adaptions are made based on the available system inputs. Usually, a decision is taken and communicated, but sometimes only the indicators (input parameters) are communicated and the decision is taken by the receiving node. In most cases, the decision involves switching to pre-defined modes (profiles) based on certain criteria, but in-mission optimizations have also been reported.

Index

© The Author(s) 2020
D. Sotnik et al. (eds.), *Cognitive Underwater Acoustic Networking Techniques*,
SpringerBriefs in Electrical and Computer Engineering,
https://doi.org/10.1007/978-3-662-61658-1

Printed in the United States
By Bookmasters